Theory of Knowledge
and the Rise of Modern Science

Theory of Knowledge and the Rise of Modern Science

Clare Hay

The Lutterworth Press

The Lutterworth Press
P.O. Box 60
Cambridge
CB1 2NT
United Kingdom

www.lutterworth.com
publishing@lutterworth.com

ISBN: 978 07188 3089 2

British Library Cataloguing in Publication Data
A record is available from the British Library

First Published in 2009

Typeset by Clare Hay System LaTeX 2_ε

Contents

Chapter 1

Introduction

Traditionally epistemology, the branch of philosophy concerned with knowledge—what knowledge is and how it is attained—begins with the question, what can I know? The obvious response is to carry out a survey of what I, or you, or whoever is carrying out the inquiry, have in mind. We carry out an inventory of our mental contents, and then we might try to establish which items are true, which ones are likely, which ones seem unlikely or improbable and which, on reflection, seem to be just plain wrong. We might try and establish a set of criteria for evaluating mental contents; some things seem to be obvious or self-evident, others plausible, and so on.

The difficulty that soon emerges is that we seem to know more than we have plausible grounds for, and this is where epistemology as a classic set of philosophical problems gets its bite. Perhaps the mind is a blank slate at birth, and all that I know I have come to know through my senses. The contents of my mind got there by sight, sound, touch, taste or smell. If so, the senses deliver to my mind what is given to them by the outer world, and my mind contains representations (literally "re-presentations") of what is outside me. The mind takes these representations and compares and contrasts them; I note that the sky is blue, the sea is blue, that man over there is wearing a blue shirt, and from these and other experiences I abstract out an idea of blueness, an idea that I can apply generally. Such an approach is classical *empiricism*, and the sorts of capacities allowed to the mind of comparing and contrasting to abstract out general ideas is a commonplace of empiricism. But it falls short when it comes to ideas that do not seem to have any ground in experience, ideas concerning God, the infinite, cause, necessity and personal identity, most of which we will touch on at some point in the following.

The other great tradition in theory of knowledge is *rationalism*. Philosophers in this tradition tend to start with a rather more elevated view of human nature. Whereas empiricists tend to take a hard-headed and, literally, down-to-earth view of what we can and cannot know, rationalists tend to see us as a mix of the worldly and the divine, with the mind as a "divine spark", a fragment of man that is at least on the way to being God-like. Knowledge comes primarily from reflection on the mind itself, which is far more than a blank slate. For the rationalist what is most important in our knowledge comes from such reflection, and this is how we get knowledge of what seems to lie beyond presentation—God, the infinite, cause, necessity, personal identity. But while the empiricist has problems with accounting for what we think we know even though it cannot be derived from the senses, the rationalist faces problems with explaining how what is primarily derived from reflecting on the mind is fitted to navigating the physical world we interact with. All too often the rationalist suggests that we have knowledge of what we cannot sensibly seem to know; if the infinite really is infinite, after all, and we are finite, how can we talk sensibly about the infinite? It must surely be beyond us.

This is the classic debate in epistemology, and it is the primary concern in the following. It can be traced back to arguments that raged in the Academy in ancient Athens, between Plato and his rationalist followers and the more empirically-minded Aristotelians. It provides the framework for the debate between the rationalists of the modern period (Descartes, Spinoza, Leibniz) and the slightly later empiricists (Locke, Berkeley, Hume). It is the basis for Kant's great synthetic works, in which he tries to reconcile these two approaches and take the best from each. But Kant's synthesis of these two great traditions led to a whole series of further arguments concerning the relations of epistemology and science.

Although our topic is epistemology, what has been said so far has touched on nearly all the mainstream topics of philosophy. We have touched on the philosophy of mind and on ontology, the study of being, of what there actually is, including the question of whether or not there is a God, and what we can know about God. We have also touched on the philosophy of science and the relations between philosophy and science. There are close interconnections between all the main branches of philosophy and it is hard to have a view in any particular branch, including epistemology, that does not have consequences for other aspects of the subject. Superficially it looks like the rationalist holds all the cards, as she can have views on God, on the mind, on our capacities to know truth, that are not obviously available to the empiricist. But on the other hand the empiricist can argue that much of this is illusory, mere speculation about what we cannot really know. And here the empiricist can point to the advances of modern science, while conveniently forgetting that both Newton and Einstein relied on thought experiments that go well beyond what can be established by means of the senses alone. At bottom philosophy does not have neat divisions, because there are all sorts of interconnections between its various branches.

As well as these relations between its own branches, philosophy has lots of interconnections with other disciplines. It borders on psychology, theology, and the social and the natural sciences. It is only the last of these that we will be concerned with. For much of the last 2500 years what we would now call "science" was called "natural philosophy", as opposed to moral philosophy. This has important consequences for reading most of the philosophers discussed here, certainly up to and including Kant, because none of them thinks that there is a significant difference between "doing philosophy" and "doing science". Although there is some divergence after Newton, which is why I have included Newton in the text, the separation of philosophy and science that we now have is one that only became established in the nineteenth century. I think you will find it helpful when reading Kant and his predecessors to think of them as contributing to science as much as to philosophy. This is certainly the line I have followed here.

This said, although the rationalism/empiricism framework is helpful, it is only a guide. It was established by Kant, in the final section of his monumental *Critique of Pure Reason*, but in many ways it reflects Kant's compulsive need to impose a systematic framework on everything he came into intellectual contact with. While it reflects broad tendencies among the people he was writing about, they rarely fit comfortably into their Kantian pigeon-holes. The framework has its value but I would encourage you to approach any of the great, and the not-so-great, philosophers with an open mind, and be prepared to be surprised. Berkeley, for example, does not fit very comfortably into any mould.

Concerning the organisation of the text, I have followed a traditional pattern of treating rationalism first, then empiricism, then Kant. This reflects a more-or-less chronological pattern, and that each of these thinkers read, or was at least aware of, the views of their contemporaries and predecessors. There was a network of intellectuals in seventeenth and eighteenth century Europe, aware of one anothers' work via correspondence, learned societies and journals as well as through published books. This network expanded greatly in the nineteenth and twentieth centuries and became increasingly university-based. After Kant I have chosen to fast-forward to the twentieth century and take a more topic-based approach.

With the best will in the world the following should not all be obvious on first reading. Unlike some disciplines philosophy does not have a linear structure. Students and practitioners of philosophy read the same great classic texts over and over, trying to understand them. When he read Descartes' *Geometry*, Newton said that he read a few pages until he did not understand any more, then he would go away for a while and let what he had read run around in his mind. He returned over and over again to the book until he had worked right through it. Wittgenstein has a wonderful metaphor (in *On Certainty*) of light dawning slowly over a plane. This is a common experience of coming to philosophy, so do not be discouraged if some of the following seems baffling on a first, or even a second or third reading. Come back later and you will often find that what seemed obscure has become clear. Studying philosophy can be the most extraordinary intellectual journey and I hope this text will give you some signposts on the way.

Turning from the sublime to the practical, I have written this book with two audiences in mind. The first is students of philosophy, from AS/Baccalaureate through to advanced-level undergraduates. While the book is certainly demanding of an AS level student, it covers the whole of the syllabus of the AQA AS module 1 on theory of knowledge. My companion book *The Theory of Knowledge, a Coursebook* (Lutterworth Press, 2008) is aimed directly at the AQA AS module 1, while this book is intended primarily for teachers of that course and more advanced students of philosophy.

The second audience is the general reader, and I have tried to write an accessible book that deals with the subject in a thorough fashion but without assuming any prior knowledge of philosophy.

I should point out that the approach taken here is distinctly unconventional. Although there are signs that Newton is edging into the mainstream philosophical canon, my approach is more heavily weighted to correlating philosophical developments and the rise of modern science than is usual in overtly philosophical introductions. Hence the quotations taken from original sources are more extensive than is usual for a book of this kind. In a manner of speaking the quotations are the evidence for a theory of the development of modern philosophy, and they are intended to be sufficient to substantiate this approach.

References are, wherever possible, to the page or section numbering used in standard editions; in the case of Plato, Stephanus page numbering, Descartes, volume/page numbering in the Adam & Tannery edition, and so on. The editions used are listed in the Bibliography. Page references for Newton are to *Principia Mathematica* and *Opticks* first, and to *Isaac Newton, Philosophical Writings* second. For *Principia Mathematica* I have used Andrew Motte's translation, but the page numbers given are for the Cohen/Whitman translation (as used in *Isaac Newton, Philosophical Writings*). Translations from Descartes' *Meditations* are by John Veitch (1901), available on-line (http://www.wright.edu/cola/-descartes/mede.html), with minor alterations. Words in **boldface** have an entry in the Glossary. These entries either refer back to the text or expand on it, and they provide an alternative way of reading the book.

Finally, a comment on my liberal use of the "vertical pronoun". The vertical pronoun is widely frowned on in educational and academic contexts, but if you look closely most authors substitute "here" or "in the present work" or some such self-referential indicator, because books and journal articles are the work of engaged, sentient beings, not robots or monkeys with typewriters. Philosophy may sometimes be about facts but it is not itself a matter of stating and arranging facts, and eliminating the vertical pronoun is, to my mind, a curiously self-defeating denial that the best philosophical works are the products of interestingly cantankerous, passionate and often partisan authors.

Relatedly when an author uses "I" they often intend the I to be you; that you are to put yourself in their shoes in order to see what they see. The I of Descartes' *Meditations*, for example, is just such an invitation to you to be I. The truly great philosophers are an amalgam of hard-headed rational, logical thinking and extraordinary insight into the human situation. You can do logic in the third person, insight is more personal and immediate.

Chapter 2

Plato's *Meno*

The first great philosopher in the Western tradition is Socrates, but Socrates did not write anything down. His disciple and successor, Plato, wrote down and expanded on Socrates' thought in a series of dialogues, the earlier of which owe a great deal to Socrates, the latter reflecting the flower of Plato's own philosophical genius. The dialogues are often literary masterpieces as well as acute philosophical analyses.

The four dialogues we will be concerned with are the *Meno*, the *Theaetetus*, the *Phaedo*, and the *Republic*, all from Plato's middle period, ca. 388–367BC. In these Plato sets out views on knowledge and what it is to attain knowledge, views that profoundly influenced his successors. The conclusions, such as they are, are mostly negative. In the dialogues Socrates engages in conversation with other Greek citizens who usually profess knowledge of a particular topic; virtue, courage, piety, justice, goodness, and so on. Socrates says that he does not know what these concepts are and asks for an account so that he, too, might grasp the concept in question. Invariably it turns out that Socrates' interlocutor comes to realise that, pressed to give an account of what they think they know, it turns out that they do not know either. Meno (in the *Meno*) compares Socrates to a jellyfish that stuns and renders speechless whoever it stings, and Anytus, an older man that Socrates involves in the dialogue, becomes angry when Socrates points out that many virtuous fathers have wicked sons, implying that the fathers do not know what virtue is because they are unable to teach it to their sons. (The exclusive concentration on the male gender is a reflection of the nature of the society of the Athens of Socrates and Plato.) Plato is deliberately alluding to the fact that shortly afterwards the Athenians condemned and executed Socrates for "corrupting the young", that is, for fearlessly pursuing intellectual inquiry into political and ethical concepts even though such inquiries threatened the authority of the entrenched political and social order. At best authority is ambivalent towards philosophy.

A major aspect of Socrates' approach is to encourage people to think through the issues themselves, and the inquiry itself is generally more valuable than any results it may yield. Socrates begins by asking for a definition, and the ensuing argument takes the form of "dialectic"—an argument in which at each stage a definition is proposed, evaluated, and found inadequate. Each successive stage builds on what has gone before. Is it possible, though, to give a definition of a concept by itself, one that does not involve other concepts? Can a definition be given of knowledge, for example, that does not refer to such concepts as belief, opinion, judgment, truth, and so on? If so, whichever concept is chosen can be taken as fundamental, and the others defined in terms of it.

It could be, though, that no single concept has such a fundamental role, that our concepts form a mutually self-supporting structure. Approaches of the former kind are thought of as **foundational**, because they assume that there is some original point on which the body of our knowledge rests. Approaches of the latter kind are known as **coherentist** because they deny foundational approaches and argue that what we know is a set of truths or beliefs that all hang together or cohere with one another. To forestall your objection, they are true, or believed, *because* they hang together—it is not that they hang together because they are true, or because they are believed. Because then truth, or belief, would be our fundamental concept. Mental alertness is part of the discipline of philosophy. As for foundationalism and coherentism, we will return to both of these later.

The second issue concerns the sorts of concepts that philosophy deals with. Can one know what knowledge is? "Knowledge" does not seem to be the sort of thing that can be known, in the sense that you can know what colour the sky is or what you had for breakfast this morning. Instances of knowledge seem to be fairly straightforward but saying what it is to know as such seems altogether more elusive. Philosophical concepts tend to have this slippery sort of self-reference and it is not always entirely clear that what we say about these concepts even makes sense. Writing philosophy is often a challenge to say what seems to be almost unsayable, and reading it can be almost as challenging.

In the first instance the *Meno* is a discussion between Socrates and Meno concerning virtue; in the main, though, we will be concerned with Socrates' conversation with Meno's slave-boy, in which Socrates draws out of the boy a knowledge of geometry that the boy did not seem to have beforehand. This part of the dialogue is intended to establish certain points, so some context is needed to appreciate what Socrates is trying to show.

2.1 Looking for what we do not know

The dialogue opens with Meno asking Socrates "whether virtue is acquired by teaching or by practice; or if neither by teaching nor practice, then whether it comes to man by nature, or in what other way?" Socrates replies that he does not know what virtue is, and asks Meno for a definition. This is duly offered and for the first third of the dialogue Socrates shows that none of Meno's proposals will work, reducing him to the state of intellectual bemusement of one stung by a jellyfish. At this point Socrates suggests that they continue the inquiry nevertheless, to which Meno replies:

Meno How will you inquire, Socrates, into that which you do not know? Where can you find a starting-point in the region of the unknown? And if you find what you want, how will you ever know that this is the thing which you did not know? (*Meno*, p.80d)

This poses a real problem for Socrates. If someone says "observe", for example, the only reply is "observe what?" Similarly you cannot just inquire, there has to be something you inquire into or after. But how can anyone ever find out or think up or discover anything new? Socrates' curious response is to offer a myth, a "glorious truth" that he has learned from priests and priestesses, of reincarnation and remembrance from past lives:

> They say that the soul of man is immortal, and at one time has an end, which is termed dying, and at another time is born again, but is never destroyed. And the moral is, that

> a man ought to live always in perfect holiness. "For in the ninth year Persephone sends the souls of those from whom she has received the penalty of ancient crime back again from beneath into the light of the sun above, and these are they who become noble kings and mighty men and great in wisdom and are called saintly heroes in after ages." The soul, then, as being immortal, and having been born again many times, and having seen all things that exist, whether in this world or in the world below, has knowledge of them all; and it is no wonder that she should be able to call to remembrance all that she ever knew about virtue, and about everything; for as all nature is akin, and the soul has learned all things; there is no difficulty in a man eliciting out of a single recollection all the rest—the process generally called "learning"—if he is strenuous and does not faint; for all inquiry and all learning is but recollection. (*Meno*, p.81b)

This sets the scene for Socrates' interchange with the slave-boy, but leaves us with several questions. In particular, how does this answer Meno's question, and why does Socrates offer it as a myth or legend, putting some distance between the tale and himself, as if he does not really believe it? Clearly Socrates is under pressure, because if acquiring knowledge involves being taught then at some point in the past someone must have discovered for themselves what is known, with no external prompting or teaching. But if this is so then they could only ever have opinion, they can never attain secure knowledge (this is bound up with the theory of forms, the topic of Chapter 4). The doctrine of recollection is a way out of this difficulty, and Socrates' diffidence about it perhaps reflects its rather unattractive character. It is certainly curious that denying reincarnation entails denying the possibility of knowledge as opposed to mere opinion.

Socrates Attend now to the questions which I ask him, and observe whether he learns of me or only remembers.

Meno I will.

Socrates Tell me, boy, do you know that a figure like this is a square?

Boy I do.

Socrates And you know that a square figure has these four lines equal?

Boy Certainly ...

Socrates A square may be of any size?

Boy Certainly.

Socrates And if one side of the figure be of two feet, and the other side be of two feet, how much will the whole be? Let me explain: if in one direction the space was of two feet, and in other direction of one foot, the whole would be of two feet taken once?

Boy Yes.

Socrates But since this side is also of two feet, there are twice two feet?

Boy There are.

Socrates Then the square is of twice two feet?

Boy Yes.

Socrates And how many are twice two feet? Count and tell me.

Boy Four, Socrates.

Socrates And might there not be another square twice as large as this, and having like this the lines equal?

Boy Yes.

Socrates And of how many feet will that be?

Boy Of eight feet.

Socrates And now try and tell me the length of the line which forms the side of that double square: this is two feet—what will that be?

Boy Clearly, Socrates, it will be double. [I.e. each side is four feet, which will give an area of 16 square feet.]

Socrates Do you observe, Meno, that I am not teaching the boy anything, but only asking him questions; and now he fancies that he knows how long a line is necessary in order to produce a figure of eight square feet; does he not?

Meno Yes.

Socrates And does he really know?

Meno Certainly not.

Socrates He only guesses that because the square is double, the line is double.

Meno True.

Socrates Observe him while he recalls the steps in regular order. (To the Boy.) Tell me, boy, do you assert that a double space comes from a double line? Remember that I am not speaking of an oblong, but of a figure equal every way, and twice the size of this—that is to say of eight feet; and I want to know whether you still say that a double square comes from double line?

Boy Yes.

Socrates But does not this line become doubled if we add another such line here?

Boy Certainly.

Socrates And four such lines will make a space containing eight feet?

Boy Yes.

Socrates Let us describe such a figure [i.e. four foot by four foot]: Would you not say that this is the figure of eight feet?

Boy Yes.

Socrates And are there not these four divisions in the figure, each of which is equal to the figure of four feet? [I.e. four squares each four feet square.]

Boy True.

Socrates And is not that four times four?

Boy Certainly.

Socrates And four times is not double?

Boy No, indeed.

Socrates But how much?

Boy Four times as much.

Socrates Therefore the double line, boy, has given a space, not twice, but four times as much.

Boy True.

Socrates Four times four are sixteen—are they not?

Boy Yes.

Socrates What line would give you a space of eight feet, as this gives one of sixteen feet;—do you see?

Boy Yes.

Socrates And the space of four feet is made from this half line?

Boy Yes.

Socrates Good; and is not a space of eight feet twice the size of this, and half the size of the other?

Boy Certainly.

Socrates Such a space, then, will be made out of a line greater than this one, and less than that one?

Boy Yes; I think so.

Socrates Very good; I like to hear you say what you think. And now tell me, is not this a line of two feet and that of four?

Boy Yes.

Socrates Then the line which forms the side of eight feet ought to be more than this line of two feet, and less than the other of four feet?

Boy It ought.

Socrates Try and see if you can tell me how much it will be.

Boy Three feet.

Socrates Then if we add a half to this line of two, that will be the line of three. Here are two and there is one; and on the other side, here are two also and there is one: and that makes the figure of which you speak?

Boy Yes.

Socrates But if there are three feet this way and three feet that way, the whole space will be three times three feet?

Boy That is evident.

Socrates And how much are three times three feet?

Boy Nine.

Socrates And how much is the double of four?

Boy Eight.

Socrates Then the figure of eight is not made out of a line of three?

Boy No.

Socrates But from what line?—tell me exactly; and if you would rather not reckon, try and show me the line.

Boy Indeed, Socrates, I do not know.

Socrates Do you see, Meno, what advances he has made in his power of recollection? He did not know at first, and he does not know now, what is the side of a figure of eight feet: but then he thought that he knew, and answered confidently as if he knew, and had no difficulty; now he has a difficulty, and neither knows nor fancies that he knows.

Meno True.

Socrates Is he not better off in knowing his ignorance?

Meno I think that he is.

Socrates If we have made him doubt, and given him the "jelly-fish's shock", have we done him any harm?

Meno I think not.

Socrates We have, it would seem, certainly assisted him in some degree to the discovery of the truth; and now he will wish to remedy his ignorance, but then he would have been ready to tell all the world again and again that the double space should have a double side.

Meno True.

Socrates But do you suppose that he would ever have inquired into or learned what he fancied that he knew, though he was really ignorant of it, until he had fallen into perplexity under the idea that he did not know, and had desired to know?

Meno I think not, Socrates.

Socrates Then he was the better for the jellyfish's touch?

Meno I think so.

Socrates Mark now the farther development. I shall only ask him, and not teach him, and he shall share the inquiry with me: and do you watch and see if you find me telling or explaining anything to him, instead of eliciting his opinion. Tell me, boy, is not this a square of four feet which I have drawn?

Boy Yes.

Socrates And now I add another square equal to the former one?

Boy Yes.

Socrates And a third, which is equal to either of them?

Boy Yes.

Socrates Suppose that we fill up the vacant corner?

Boy Very good.

Socrates Here, then, there are four equal spaces?

Boy Yes.

Socrates And how many times larger is this space than this other?

Boy Four times.

Socrates But it ought to have been twice only, as you will remember.

Boy True.

Socrates And does not this line, reaching from corner to corner, bisect each of these spaces?

Boy Yes.

Socrates And are there not here four equal lines which contain this space?

Boy There are.

Socrates Look and see how much this space is.

Boy I do not understand.
Socrates Has not each interior line cut off half of the four spaces?
Boy Yes.
Socrates And how many spaces are there in this section?
Boy Four.
Socrates And how many in this?
Boy Two.
Socrates And four is how many times two?
Boy Twice.
Socrates And this space is of how many feet?
Boy Of eight feet.
Socrates And from what line do you get this figure?
Boy From this.
Socrates That is, from the line which extends from corner to corner of the figure of four feet?
Boy Yes.
Socrates And that is the line which the learned call the "diagonal". And if this is the proper name, then you, Meno's slave, are prepared to affirm that the double space is the square of the diagonal?
Boy Certainly, Socrates.

Diagramatically, this is how the proof works. The slave-boy begins by drawing a square with sides of 2 feet:

When asked to double this, he doubles the length of the side:

Which, as Socrates points out, is not twice the area but four times (16 square feet rather than 8). So Socrates suggests drawing a square as first drawn, and adding to it:

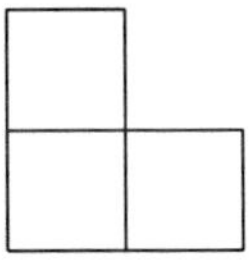

Filling in the last corner gives the slave-boy's first doubled figure, but this time Socrates draws in the diagonals:

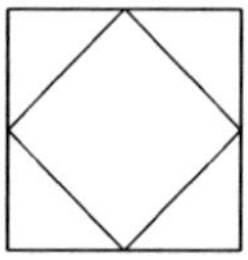

This gives four-right angled triangles, each like this:

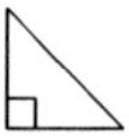

The area of these is $\frac{1}{2}$ base $\times$ height, which is $\frac{1}{2} \times 2 \times 2 = 2$ $\times 4$ (4 trianges) $= 8$. $16 - 8$ (the area of four triangles, each two units square) $= 8$; so putting in the diagonals yields a square that is indeed twice the area, as intended. It might look like the diagrams are an essential part of the proof, but they are not. These diagrams were drawn using a system of Cartesian co-ordinates, so it is possible to produce an algebraic proof that does not rely on visual intuition at all.

It can also be proved using Pythagoras' theorem, that the square of the hypotenuse (long side) of a right-angled triangle is equal to the squares of the two short sides added together. Our triangle has sides of length 2, so the length of the hypotenuse is $\sqrt{2^2 + 2^2} = \sqrt{4+4} = \sqrt{8}$. This is the length of one side of the square, and since the area of a square is the length of the side squared, the area $= \sqrt{8} \times \sqrt{8} = 8$, which is what was to be shown, so in conventional fashion we will end our proof ... $\square$

Socrates What do you say of him, Meno? Were not all these answers given out of his own head?

Meno Yes, they were all his own.

Socrates And yet, as we were just now saying, he did not know?

Meno True.

Socrates But still he had in him those notions of his—had he not?

Meno Yes.

Socrates Then he who does not know may still have true notions of that which he does not know?

Meno He has.

Socrates And at present these notions have just been stirred up in him, as in a dream; but if he were frequently asked the same questions, in different forms, he would know as well as anyone at last?

Meno I dare say.

Socrates Without anyone teaching him he will recover his knowledge for himself, if he is only asked questions?

Meno Yes.

Socrates And this spontaneous recovery of knowledge in him is recollection?

Meno True.

Socrates And this knowledge which he now has must he not either have acquired or always possessed?

Meno Yes.

Socrates But if he always possessed this knowledge he would always have known; or if he has acquired the knowledge he could not have acquired it in this life, unless he has been taught geometry; for he may be made to do the same with all geometry and every other branch of knowledge. Now, has any one ever taught him all this? You must know about him, if, as you say, he was born and bred in your house.

Meno And I am certain that no one ever did teach him.

Socrates And yet he has the knowledge?

Meno The fact, Socrates, is undeniable.

Socrates But if he did not acquire the knowledge in this life, then he must have had and learned it at some other time?

Meno Clearly he must.

Socrates Which must have been the time when he was not a man?

Meno Yes.

Socrates And if there have been always true thoughts in him, both at the time when he was and was not a man, which only need to be awakened into knowledge by putting questions to him, his soul must have always possessed this knowledge, for he always either was or was not a man?

Meno Obviously.

Socrates And if the truth of all things always exists in the soul, then the soul is immortal. Wherefore be of good cheer, and try to discover—by recollection—what you do not now know, or rather what you do not remember.

Meno I feel, somehow, that I like what you are saying. (*Meno*, p.82b–86b)

However much Meno might like what Socrates is saying, it soon becomes clear in the dialogue that what works for a geometrical proof does not readily carry over into an inquiry into virtue. Is it plausible that the slave-boy could have constructed the proof if Socrates had done no more than point out where he was going astray? Is it really the case that in his questioning he does not "lead the witness"? After all the key point Socrates makes to Meno early on is that "I literally do not know what virtue is," so how is recollection to be brought about in the absence of a guide who knows at a conscious rather than a subconscious level? This is why Socrates ends up angering Anytus by pointing out that even those who are virtuous do not seem to know what virtue is because they are unable to teach it to their children. Anytus, himself a father, takes this is an insult to him and his father, but in truth Socrates is in danger of painting himself into a corner. If someone can be virtuous without knowing what virtue is, and surely there are virtuous people, then somehow their virtuousness must flow from something they have in mind. It cannot be a matter of chance alone that some people are virtuous and others wicked, and Anytus is right to be angered by the implication of Socrates' argument that those who are virtuous are no wiser or more enlightened than those who lie, cheat, and defraud. It cannot be the case that there is either knowledge or ignorance with no middle ground.

2.2 "Right opinion" and knowledge

This leads Socrates to distinguish "right opinion" from knowledge:

Socrates If a man knew the way to Larisa, or anywhere else, and went to the place and led others thither, would he not be a right and good guide?

Meno Certainly.

Socrates And a person who had a right opinion about the way, but had never been and did not know, might be a good guide also, might he not?

Meno Certainly.

Socrates And while he has true opinion about that which the other knows, he will be just as good a guide if he only thinks the truth, as he who knows the truth?

Meno Exactly.

Socrates Then true opinion is as good a guide to correct action as knowledge; and that was the point which we omitted in our speculation about the nature of virtue, when we said that knowledge only is the guide of right action; whereas there is also right opinion.

Meno So it seems.

Socrates Then right opinion is not less useful than knowledge?

Meno There is a difference, Socrates; he who has knowledge will always be right, but he who has right opinion will sometimes be right, and sometimes not.

Socrates What do you mean? Can he be wrong who has right opinion, so long as he has right opinion?

Meno I admit the cogency of your argument, and therefore, Socrates, I wonder that knowledge should ever be prized far above right opinion—or why they should ever differ.

Socrates And shall I explain this wonder to you?

Meno Do tell me.

Socrates You would not wonder if you had ever observed the images of Daedalus; but perhaps you have not got them in your country?

Meno What have they to do with the question?

Socrates Because they require to be fastened in order to keep them, and if they are not fastened they will run away like fugitive slaves.

Meno Well, what of that?

Socrates I mean to say that like runaway slaves they are not very valuable possessions if they are at liberty, for they will walk off; but when fastened, they are of great value, for they are really beautiful works of art. Now this is an illustration of the nature of true opinions: while they abide with us they are beautiful and fruitful, but they run away out of the human soul, and do not remain long, and therefore they are not of much value until they are fastened by the tie of the cause; and this fastening of them, friend Meno, is recollection, as you and I have agreed to call it. But when they are bound, in the first place, they have the nature of knowledge; and, in the second place, they are abiding. This is why knowledge is more honourable and excellent than true opinion, because fastened by a chain. (*Meno*, p.97a–98a)

First of all, Socrates explicitly connects the concepts of belief, truth, and knowledge. If someone says that they believe something or other, say, that the moon is made of green cheese, we naturally assume that they hold their belief to be true. This is shown by the ways in which we act on our beliefs. If I believe there is a bottle of wine in the cupboard, and I desire a glass of wine, I show that I believe this to be so—my belief to be true—by heading towards the cupboard with a corkscrew and a glass in hand. But I'm prepared to find that my belief will be proven false and that in truth there is no wine. How one is to manifest one's belief that the moon is made of green cheese, or in any other belief that cannot be readily acted on, is another matter which, for the moment, we will keep as an open question.

That beliefs may turn out to be wrong leads Meno to distinguish belief from knowledge, because "he who has knowledge will always be right; but he who has right opinion [belief] will sometimes be right, and sometimes not." Beliefs are **defeasible**, that is, we assume they are correct but are prepared to change them if the evidence goes against them. By contrast I can only be said to know if what I know is true; unlike belief, knowledge does not seem to be defeasible.

This puts a very high requirement on knowledge and, as Meno says, since right (i.e. well-founded) beliefs are good enough, it is not clear why we should want to get from such beliefs to knowledge—even if we knew how to take the step, what good would it do us?

2.3 "Fastened by the tie of the cause"

At the end of the last passage quoted above, Socrates suggests a difference between belief and knowledge. The latter, unlike the former, is "fastened by the tie of the cause." It seems that what is known—what is true—is not different in kind from what is believed, but is somehow located in a way that links up with the concept of cause. But how? In an informal sense we operate this way on an everyday basis; why do I believe that there is wine in the cupboard? Because I bought two bottles last weekend and I only recall finishing one. Why do I think that the contents of the second bottle, if it is still there, are wine? Because I have a theory about wine merchants and suppliers and viticulture and grapes, stretching right back to the sun that ripens the grapes—which incidentally furnishes me with a reason for thinking that my friend's belief that the moon is made of green cheese is wrong. If you think about almost anything that you take for granted you will find a whole series of "why?" questions with "because" answers. These causal chains comprise *theories*, where what is distinctive of a theory is that it is *systematic.* We go about things systematically in order to do whatever it is we set out to do, and if things go wrong we look for a reason—for a cause of what went wrong so that we can adjust our theory accordingly for future occurrences.

What Socrates proposes here is a *coherence* theory of knowledge (**coherentism**), one in which a sufficiently large number of beliefs hang together and which, if we act on them, result in successful outcomes. But does this really work? How extensive must a body of beliefs be for it to become knowledge? And how can we know that this "critical mass" has been achieved? It is not at all clear that Socrates has given us a basis to distinguish beliefs that work for us, beliefs that we think of as true but also as defeasible, from truths that are known and that we do not think of as defeasible. To take Socrates' approach, can you think of a "because" answer to the question "why is *this* truth (whatever it may be) an item of knowledge and not just a belief?"

Socrates comes up with an answer when he says "this fastening of [beliefs], friend Meno, is recollection, as you and I have agreed to call it," referring back to the "glorious truth" that he learned from priests and priestesses, as quoted in 2.1. But this has not been agreed on, much less explained. In truth this is one of those fascinating points, well worth looking out for in philosophical works, where the author is struggling to get their argument to hang together. Socrates wants a way of getting from belief to knowledge, but the coherence approach cannot quite do this. That beliefs do not come one by one, or at least cannot be explained or justified one by one, is shown by series of "because" responses to "why?" questions. If they could be explained or justified one by one such series would not get off the ground. But equally there is no obvious sense in which coherence gives a condition, a criterion, for distinguishing belief from knowledge.

If, then, coherence cannot do what is required, what of recollection? What would the idea that we learn geometry by a programme of guided questioning that brings about recollection, rather than what we would usually think of as teaching, add to result in knowledge rather than belief? But it is suspicious that what works for a mathematical problem is not obviously transferable to, say, a moral dilemma. Certainly most people think that mathematical knowledge is very different in kind from other sorts of knowledge, and this makes the dialogue between Socrates and the slave-boy look like a set-piece whose results are not readily applicable to other walks of life. When the slave-boy suggests that doubling the side will double the area, as an initial (false) belief, the proof gets underway; what would have followed if, say, Socrates had asked the slave-boy about the composition of the moon or if telling white lies is wrong?

The *Meno* seems, at bottom, rather inconclusive, and bringing in recollection looks like an attempt to fill a hole in the argument. The open question remains, how does knowledge differ from right belief? What additional whatever-it-is must knowledge have that right belief lacks? Given that what is involved in belief and knowledge is a relation between minds and what is external to us, which might or might not be limited to what exists in space and time, the difference could be a function either of our minds or of what there is. We may be able to identify a special mental act or criterion whose presence (or absence) enables us to distinguish knowledge from belief. Or perhaps the difference lies outside of us, and we can only have knowledge of particular sorts of things, such that beliefs can only be of other sorts of things. We will see both of these strategies invoked in the following. For the moment, even if we do not have any answers, at least the questions are a bit clearer.

Chapter 3

Plato's *Theaetetus*

3.1 Socrates the midwife

In the Introduction a main difficulty for theory of knowledge was proposed—that we think we know, or we seem to know, more than we have any straightforward (i.e. perceptual) basis for knowing. We can take it for read that to have knowledge is also to have a grasp of truth. To have knowledge is to know what is true; if I think I know something but whatever-it-is that I think I know is false, I do not know a falsehood. In such a case I lack knowledge—I simply do not know. I can know that such-and-such is not the case, but what I know is a truth—that such-and-such is not the case. This is why knowledge is bound up with truth and this marks the difference between knowledge and belief, because if I have a false belief then I do not lack belief, what I have in mind is simply false. It makes sense to say, I used to believe that such-and-such is the case, but now I know better, whereas it makes no sense to say I used to know that such-and-such is the case, but now I know better.

In the previous chapter on the *Meno* two ways of distinguishing belief from knowledge were discussed: knowledge is either a special sort of mental act whose objects (truths) have a special, identifiable mental criterion associated with them, or knowledge is of special sorts of objects, where these objects are external to us. Plato considers both of these. The first is the subject of the *Theaetetus*, a long and difficult dialogue that we will touch on in this chapter. The second is perhaps Plato's most famous doctrine, the theory of forms, which we will discuss in the next chapter (on the *Phaedo* and the *Republic*).

Put like this, it seems that Plato proposes two incompatible approaches to knowledge. But this is to assume that one approach is right and the other wrong and, as it is knowledge that we are after, that there is an argument or proof to establish as a matter of knowledge that one is right and the other wrong. It does not actually work like this, and this reflects one of the strengths of Plato's way of presenting philosophy. The dialogue form goes with a flexible, dynamic and open-ended approach, in which objections can be raised and discussed. And like most arguments the dialogues are rarely if ever conclusive. The dialogue form enables Plato to discuss different points of view, to encourage his reader to think through the cut-and-thrust of the debate. The dialogues are intended to engage, not to teach, in no small measure because Socrates claims that he lacks knowledge and consequently has nothing to teach. In the *Theaetetus* he likens his art to that of the midwife:

> My art of midwifery is in most respects like theirs; but differs, in that I attend men and not women; and look after their souls when they are in labour, and not after their bodies: and the triumph of my art is in thoroughly examining whether the thought which the mind of the young man brings forth is a false idol or a noble and true birth. And like the midwives, I am barren, and the reproach which is often made against me, that I ask questions of others and have not the wit to answer them myself, is very just—the reason is, that the god compels me to be a midwife, but does not allow me to bring forth. And therefore I am not myself at all wise, nor have I anything to show which is the invention or birth of my own soul, but those who converse with me profit. Some of them appear dull enough at first, but afterwards, as our acquaintance ripens, if the god is gracious to them, they all make astonishing progress; and this in the opinion of others as well as in their own. It is quite clear that they never learned anything from me; the many fine discoveries to which they cling are of their own making. But to me and the god they owe their delivery. (*Theaetetus*, p.150b–d)

This fits with the suggestion in the *Meno* that knowledge is recollection (p.10, above), but it is not clear that it evades the paradox Meno points out to Socrates, that you can hardly search or inquire when you do not know what it is you are searching for or inquiring after. But at bottom it is hard to deny that we have some idea of what knowledge is. The account we saw Socrates give in the *Meno* is that knowledge is true opinion "fastened by the tie of the cause", that an item of knowledge is such because it has a place in a systematic body of knowledge. Coherence provides an explanation of, and the justification for, holding that a given belief counts as knowledge. Noting that while this gives a method of getting from belief to knowledge in terms of our thinking about what there is (what we have knowledge of), it does not, so far at least, set any conditions on what it is that we have knowledge of. For the moment our question is what it is to have knowledge, irrespective of what that knowledge is of.

3.2 Tripartite definition of knowledge

In the dialogue Theaetetus expresses the view that even if the belief you hold happens to be true, if you cannot give an explanation of it, it cannot be considered to be knowledge:

> There is a distinction, Socrates, which I have heard made by someone else, but I had forgotten it. He said that true opinion, combined with reason, was knowledge, but that the opinion which had no reason was out of the sphere of knowledge; and that things of which there is no rational account are not knowable—such was the singular expression which he used—and that things which have a reason or explanation are knowable. (*Theaetetus*, p.201c–d)

This is known as the "tripartite definition of knowledge", that knowledge is justified true belief. The idea is that this is a definition, and it serves as an *analysis* of the concept. In other words, on the left-hand side we have the concept that we want to get a better understanding of, and on the right-hand side we have a set of conditions that unpack fully the target concept. If these conditions are ones that we already understand then, in theory, we have a perspicuous analysis of our target concept in simpler and readily understandable terms.

For this to work the conditions on the right-hand side have to be **necessary and sufficient**. The idea is that whatever meets the conditions set out on the right-hand side has to be an instance of the concept on the left-hand side. For something to be scarlet, for example, it has necessarily to be red. Nothing can be scarlet without being red. But being red is insufficient for instancing the concept 'scarlet'; to be scarlet something must be a particular shade of red. The underlying idea is that if knowledge is justified true belief, then whatever is justified, true and believed, is an instance of knowledge. If all goes to plan the analysis captures our target concept in a way that enables us to get a much better understanding of it.

In the present case, though, difficulties arise with the concept of justification. Intuitively we have a pretty good grasp of what it is to hold a belief; beliefs are, after all, what we act on, what we manifest a commitment to. **Truth** again we have a good intuitive grasp on. Here is one of the most famous definitions in philosophy, from Book Gamma (p.1011b) of Aristotle's *Metaphysics*: "falsity is the assertion that that which is is not, or that that which is not is and truth is the assertion that that which is is and that that which is not is not." The idea that knowledge is justified true belief sounds promising, but what counts as an adequate justification is really quite problematic. Say, for example, your friend agrees to pick you up in town and you believe that he drives a green Ford even though you have never actually seen his car. He duly picks you up in a green Ford, on which basis you think you now know that he drives a green Ford. All the evidence is in favour and it certainly appears that you have reasons, justification, for this piece of knowledge.

In fact, though, his green Ford is being repaired and the garage lent him another car, also a green Ford, for the day. Is your knowledge that he drives a green Ford well founded? (Problems of this kind are known as "Gettier problems", after a well-known paper by Edmund Gettier (Gettier 1963).) How would you adjust your beliefs—what you thought you knew—if you said "your car is rather grubby" and he pointed out that it is not his car but the garage's loan car? And if, as in this case, what seems to be a well-founded belief does indeed turn out to be wrong, is not there a risk that the conditions for what counts as knowledge are so high that it is improbable that *anything* could meet these standards?

The difficulty arises from connections between the concepts of truth and justification. To the question, why did you think that your friend drives a green Ford?, you would respond with a set of *reasons.* It is central to being able to justify a belief to be able to give an account of why you hold it. If someone says "I believe that *Dark Matter* will win the 2.40 at Newmarket tomorrow," and you bet £100 at 10:1 on *Dark Matter*, and he strolls in last, you would probably ask your friend why she tipped that horse. If she replies "no reason at all" or "I put a pin in the paper and got that name" you're likely to be pretty annoyed, as she has not really offered any sort of reason or justification at all. In court cases a distinction is sometimes made between a defence that fails, and no defence at all; putting a pin in a list is not a defence that fails, it is not a defence at all. It seems reasonable to claim that the concept of justification is bound up with those of *belief* and *reason*, with what it is to be *rational.* The rational agent is one who can offer a coherent, reasoned defence of what they believe to be the case.

Given that we have established a connection between belief and justification, how does *truth* fit into the equation? This is where it gets a bit more difficult, because there are (at least) two different ways of thinking about truth, termed "realist" and "epistemic" respectively.

A **realist** conception of truth holds that what is true, is true entirely independently of anyone judging, thinking, or believing it to be true. If it is true that Alpha Centauri is a desolate wasteland then it is true whether any human being ever gets there to find out. By contrast epistemic approaches to truth connect it with our capacities to ascertain whether or not a given state of affairs is the case. On this view the state of the surface of Alpha Centauri may or may not be a desolate wasteland, but the sentence "The surface of Alpha Centauri is a desolate wasteland" can only be said to be true (or shown to be false) when someone gets in a position to find out one way or the other. Epistemic approaches to truth are consequently **anti-realist**.

Let's say that knowledge is justified true belief, and see how this varies depending on which conception of "true" we have in mind. That your friend drives a green Ford is true if he drives a green Ford. How do you find this out? You go round to his house, look in the garage, ring him up and ask for a lift and see what he turns up in, ask the neighbours, and so on. Say you collect the evidence and propose, on this basis, that "my friend drives a green Ford" is true, because ...(insert evidence here). But clearly you have established this truth in terms of the epistemic conception; consequently there is a direct connection between truth in the epistemic sense and justification.

But it may be that your friend drives a green Ford, irrespective of your sleuthing. That "your friend drives a green Ford" is true in the realist sense. Given this, it is far from clear that there is any connection between the truth of this sentence and any justification you might give. At bottom if you hold a belief to be true in terms of justified true$_e$ (true in the epistemic sense) belief, then (as the Gettier problem shows) it could be that your belief is false. Because your belief is **defeasible** there is no way that you can claim *knowledge*. But if you assert a belief to be true in terms of justified true$_r$ (true in the realist sense) belief, then there is no connection between truth and justification; so whatever knowledge is it cannot be justified true$_r$ belief.

This is often illustrated by talking about sentences for which we have no evidence, because they fall into one of these three classes:

1. they are about the distant, undocumented past,
2. they are about inaccessible regions of space-time,
3. they express infinitely complex problems (for example, whether the expansion of π will ever generate seven consecutive 7s).

Is it true, for example, that Caesar had a bad cold on the day he crossed the Rubicon in 49BC? In the absence of evidence, while it seems that this must be true$_r$ or false$_r$ there is no obvious sense in which we can say it is true$_e$ or false$_e$. Such cases bring out the connection between truth and justification. If truth is truth$_r$, then what is true, and equally what is false, is so irrespective of us. There is nothing for the concept of justification to get a grip on. The only available answer to the question "how do you know that *this* is true?" is "it just is". It is flat-out true$_r$ and that is all there is to say about it. Justification has dropped out of the frame.

On the other hand if our concept of truth is truth$_e$, the answer to the question, "how do you know that *this* is true?" is "because x, y, and z", where x, y and z taken together constitute adequate justification. But now the concept of truth itself has dropped out of the equation. What matters is that a sentence can be justified. If it can be justified, the claim that it is true is superfluous. (True, *and* justified? What riches!) If you have a right shoe and a left shoe then you have a pair of shoes, it is not the case that you have a right shoe and a left shoe *and* a pair of shoes. The significance of the Gettier problem is a matter of ongoing debate, and an academic cottage industry has arisen in its wake. Some of the issues discussed here will also be returned to in Chapter 14.

Towards the end of the *Theaetetus*, Plato discusses three accounts of explanation, the third of which is this:

Socrates There is, further, the popular notion of telling the mark or sign of difference which distinguishes the thing in question from all others. [I.e. its unique identifying characteristic.]

Theaetetus Can you give me an example of such a definition?

Socrates As, for example, in the case of the sun, I think that you would be contented with the statement that the sun is the brightest of the heavenly bodies which revolve about the earth.

Theaetetus Certainly.

Socrates Understand why:—the reason is, as I was just now saying, that if you get at the difference and distinguishing characteristic of each thing, then, as many men affirm, you will get at the definition or explanation of it; but while you lay hold only of the common and not of the characteristic notion, you will only have the definition of those things to which this common quality belongs.

Theaetetus I understand you, and your account of definition is in my judgment correct.

Socrates But he, who having right opinion about anything, can find out the difference which distinguishes it from other things will know that of which before he had only an opinion.

Theaetetus Yes; that is what we are maintaining.

Socrates Nevertheless, Theaetetus, on a nearer view, I find myself quite disappointed; the picture, which at a distance was not so bad, has now become altogether unintelligible.

Theaetetus What do you mean?

Socrates I will endeavour to explain: I will suppose myself to have true opinion of you, and if to this I add your definition, then I have knowledge, but if not, opinion only.

Theaetetus Yes.

Socrates The definition was assumed to be the interpretation of your difference.

Theaetetus True.

Socrates But when I had only opinion, I had no conception of your distinguishing characteristics.

Theaetetus I suppose not.

Socrates Then I must have conceived of some general or common nature which no more belonged to you than to another.

Theaetetus True.

Socrates Tell me, now—how in that case could I have formed a judgment of you any more than of anyone else? Suppose that I imagine Theaetetus to be a man who has nose, eyes, and mouth, and every other member complete; how would that enable me to distinguish Theaetetus from Theodorus, or from some outer barbarian? [This is a *reductio ad absurdum*; Socrates knows he is talking to Theaetetus but according to the view on offer he can only know that he is talking to a man who bears some resemblance to Theaetetus.]

Theaetetus How could it?

Socrates Or if I had further conceived of you, not only as having nose and eyes, but as having a snub nose and prominent eyes, should I have any more notion of you than of myself and others who resemble me?

Theaetetus Certainly not.

Socrates Surely I can have no conception of Theaetetus until your snub-nosedness has left an impression on my mind different from the snub-nosedness of all others whom I have ever seen, and until your other peculiarities have a like distinctness; and so when I meet you tomorrow the right opinion will be recalled?

Theaetetus Most true.

Socrates Then right opinion implies the perception of differences?

Theaetetus Clearly.

Socrates What now shall we say of adding reason or explanation to right opinion? If the meaning is, that we should form an opinion of the way in which something differs from another thing, the proposal is ridiculous.

Theaetetus How so?

Socrates We are supposed to acquire a right opinion of the differences which distinguish one thing from another when we have already a right opinion of them, and so we go round and round:—the revolution of the scytal, or pestle, or any other rotatory machine, in the same circles, is as nothing compared with such a requirement; and we may be truly described as the blind directing the blind; for to add those things which we already have, in order that we may learn what we already think, is like a soul utterly benighted.

Theaetetus Tell me; what were you going to say just now, when you asked the question?

Socrates If, my boy, the argument, in speaking of adding the definition, had used the word "know", and not merely "have an opinion" of the difference, this which is the most promising of all the definitions of knowledge would have come to a pretty end, for to know is surely to acquire knowledge.

Theaetetus True.

Socrates And so, when the question is asked, What is knowledge? this fair argument will answer "Right opinion with knowledge",—knowledge, that is, of difference, for this, as the said argument maintains, is adding the definition.

Theaetetus That seems to be true.

Socrates But how utterly foolish, when we are asking what is knowledge, that the reply should only be, right opinion with knowledge of difference ...! (*Theaetetus*, p.208c–210a)

This is, to all intents and purposes, the end of the dialogue—an inconclusive end to its discussion of the question, "what is knowledge?" If knowledge is to know how to uniquely identify something, to know it on a plane with truth$_r$ rather than just truth$_e$, then the definition offered is circular—to know something is to know what that something is, but to know what that something is is to know it. What was wanted was a marker or criterion that would enable us to say, "this I know" as opposed to "this I believe I know but I'm not completely sure that my belief will not turn out to be false (and hence I do not know)."

The attempt to find a way of getting from belief to knowledge seems to have failed. The reason for this is the attempt to construe knowledge as a mental state out of elements including truth, when truth in a substantive sense—truth$_r$—is not a mental state at all. Any analysis of the concept of knowledge as a mental state in terms of truth is consequently doomed to fail, and this is why attempts to "solve" the Gettier problem have all, sooner or later, run into the sand. But the possibility remains open that what distinguishes knowledge from belief is not what we have in mind, or how we have it in mind, but what it is that is the *object* of belief and of knowledge. This idea can be seen at work in the *Republic* and the *Phaedo*.

Chapter 4

Plato's Theory of Forms

If knowledge involves knowledge of special sorts of things or entities, then issues in theory of knowledge shade off into issues in ontology (the study of being, of what there is) and science, because what we take there to be is usually a function of our current scientific theories. (If superstring theory best accounts for our observations of the universe, does this mean that there are superstrings?) One approach is to allow for whatever sorts of things or entities a theory proposes, and see how useful they are in explanation (how well they fit into our systematic body of theory about the world).

As a first approximation, Plato's theory of forms says that a **form** is what it is that is instantiated by two or more things that have a common property. The theory is intuitively plausible. It seems reasonable to say that there is something that all cats or all red things or all samples of gold have in common. There is a characteristic or **essence** that enables us to collect together all instances of a particular quality or property and differentiate them from instances of other qualities or properties. But if we say, as in the quote from *Theaetetus*, p.208c–210a near the end of the last chapter, that what enables us to identify red things and differentiate them from non-red things is that we know how to identify red things and differentiate them from non-red things, we do not get very far. Perhaps we should follow through this intuition and suggest that there is something, a **form**, that all instances of a particular quality or property have in common and that to have knowledge is to have knowledge of forms; where to have knowledge of a form is to have a capacity to carry out the right sort of identification and differentiation.

In the *Republic*, Plato says this:

Socrates Well then, shall we begin the inquiry at this point, following our usual method? Whenever a number of individuals have a common name, we assume that there is one corresponding idea or form:—do you understand me?

Glaucon I do.

Socrates Let us take, for our present purpose, any instance of such a group; there are beds and tables in the world—many of each, are there not?

Glaucon Yes.

Socrates But there are only two ideas or forms of such furniture; one the idea of a bed, the other of a table.

Glaucon True.

Socrates And the maker of either of them makes a bed or he makes a table for our use, in accordance with the idea—that is our way of speaking in this and similar instances—but no artificer makes the idea itself: how could he? ...God, whether from choice or necessity, made one bed in nature and one only; two or more such beds neither ever have been nor ever will be made by God.

Glaucon Why is that?

Socrates Because even if he had made but two, a third would still appear behind them of which they again both possessed the form, and that would be the real bed and not the two others.

Glaucon Very true.

Socrates God knew this, I suppose, and he desired to be the real maker of a real bed, not a carpenter making a particular bed, and therefore he created a bed which is essentially and by nature one only. (*Republic*, p.596a–b, 597c–d)

4.1 An ideal whatever-it-is

Plato's approach has been extremely influential, so much so that it underlies some of our everyday speech patterns. If someone talks of an "ideal friend" or an "ideal holiday" (even an "ideal home") we have some conception of what they have in mind. It is not that such things exist in the here-and-now, rather that they serve as yardsticks to measure against; maybe your holiday in Minorca or your friend's behaviour last night fell short of the ideal you have in mind, in much the same way as we talk of falling short of our ideals in various of our doings. And just as Meno points out in his remark to Socrates on knowledge, in 2.1, even if we did come across an ideal instance of whatever-it-is we would not know that it is, because we would not know enough to realise that it does not get any better. If, for the sake of the argument, knowledge can only be in terms of truth_r, because forms are by their very nature independent of us and our inquiries, they are the perfect objects for knowledge. Unfortunately, though, it is far from clear that we can ever do better than truth_e, which entails that we can only ever have beliefs about *instances* of forms, and hence that we can never attain knowledge (since to know is to have knowledge of a form). This being so, the reader has every right to expect of Plato an account of how we can attain knowledge of forms:

Socrates There is another thing, Simmias: is there or is there not an absolute justice?

Simmias Assuredly there is.

Socrates And an absolute beauty and absolute good?

Simmias Of course.

Socrates But did you ever behold any of them with your eyes?

Simmias Certainly not.

Socrates Or did you ever reach them with any other bodily sense? And I speak not of these alone, but of absolute greatness, and health, and strength, and in short of the reality or true nature of everything. Is the truth of them ever perceived through the bodily organs? Or rather, is not the nearest approach to the knowledge of their several natures made by him who so orders his intellectual vision as to have the most exact conception of the essence of each thing which he considers?

Simmias Certainly.

Socrates And he attains to the purest knowledge of them who goes to each with the intellect alone, not introducing or intruding in the act of thought sight or any other sense together with reason, but with the intellect in its own purity searches into the truth of each thing in its purity; he who has got rid, as far as he can, of eyes and ears and, so to speak, of the whole body, these being in his opinion distracting

> elements which when they associate with the soul hinder her from acquiring truth and knowledge—who, if not he, is likely to attain the knowledge of true being? (*Phaedo*, p.65d–66a)

Given that these ideals are not to be found in the world, we cannot find out about them by sense-perception. If they are not to be found in the world they can only be found in the mind, or at least, in a region that only the mind has access to. And this fits with the approach Socrates takes to the slave-boy in the *Meno*, that knowledge is already in the boy's head and all he has to do is prompt him to bring about recollection. It also underlies Socrates' claim in the *Theaetetus* to be a midwife who assists us in giving intellectual birth to what we already have in mind, even though we are not aware of it. Several strands of Plato's thinking come together here in a satisfying way; the theory of forms, that we have them in mind from birth because we saw them "in the world below" (*Meno*, p.81b at p.10, above), that we cannot perceive them in the world around us, and that learning consists in recollecting what is buried within our minds. Needless to say, there are problems with this. If to have knowledge is to have knowledge of the forms, and if to be ignorant is not to have knowledge at all, what is the object of belief, if belief falls somewhere between the two?

Socrates And, as knowledge corresponds to being and ignorance must obviously correspond to not-being, we have now to discover, for this intermediate between being and not-being, a corresponding intermediate between ignorance and knowledge, if there be such?

Glaucon Certainly.

Socrates Do we admit the existence of opinion?

Glaucon Undoubtedly.

Socrates As being the same faculty as knowledge, or another?

Glaucon Another.

Socrates Then opinion and knowledge have to do with different things, each according to its faculty?

Glaucon Yes.

Socrates And knowledge is relative to being [i.e. to what is, the forms] and knows being as it is ...

Glaucon Yes.

Socrates And the characteristic of opinion is to form opinions, did not we say?

Glaucon Yes.

Socrates About the same object which is known to knowledge? And will the same thing be both known and opined? Or is that not impossible?

Glaucon No, that has already been shown; if difference in faculty implies difference in subject-matter, and if, as we say, opinion and knowledge are distinct faculties, then the subject-matter of knowledge and opinion cannot be the same.

Socrates Then if being is the subject-matter of knowledge, something other than being must be the subject-matter of opinion?

Glaucon Yes, something else.

Socrates Well then, is not-being the subject-matter of opinion? Or, rather, how can there be even an opinion about that which is not? Reflect: when a man has an opinion, does he not refer it to something? Can he have an opinion which is an opinion about nothing?

Glaucon Impossible.

Socrates He who has an opinion has an opinion about some one thing.

Glaucon Yes.

Socrates And not-being is not one thing but, properly speaking, nothing?

Glaucon True.

Socrates Of not-being, ignorance was assumed to be the necessary correlative; of being, knowledge?

Glaucon And rightly.

Socrates Then opinion is not concerned either with being or with not-being?

Glaucon Not with either.

Socrates And can therefore be neither ignorance nor knowledge?

Glaucon That seems to be true.

Socrates But is opinion to be sought without and beyond either of them, in a greater clearness than knowledge, or in a greater darkness than ignorance?

Glaucon In neither.

Socrates Then I suppose that opinion appears to you to be darker than knowledge, but lighter than ignorance?

Glaucon Both [lighter and darker], and in no small degree.
Socrates And also to be within and between them?
Glaucon Yes.
Socrates Then you would infer that opinion is intermediate?
Glaucon No question.
Socrates But have we not said before, that if anything appeared to be of a sort which is and is not at the same time, that sort of thing would appear to lie in the interval between pure being and absolute not-being; and that the corresponding faculty would be neither knowledge nor ignorance, but will be found in the interval between them?
Glaucon True.
Socrates And in that interval there has now been discovered something which we call opinion?
Glaucon There has. (*Republic*, p.477a–b, 478a–d)

Knowledge, then, is of the forms, and can only be of truth (i.e. truth_r), whereas opinion is about something else; broadly what we experience in sense-perception. In the *Theaetetus* the difficulty was of getting from belief to knowledge, of what needs to be added to belief to get to knowledge. There is, though, no difference in principle between belief and knowledge in the sense that they are of the same sorts of things. In the *Meno*, the difference lies in how what is known fits together into a systematic whole (is "fastened by the tie of the cause"); what is known coheres more tightly than what is only believed. But with the theory of forms in the *Phaedo* and the *Republic*, there is a significant difference between belief and knowledge, because the objects are different in the two cases. I can have all sorts of beliefs about beds, and also, perhaps, have knowledge of the form of bed, the bed-in-itself that God created, but it is not at all clear what the one has to do with the other. In practice the implied relation between knowledge of the form of the bed and beliefs about my bed is as obscure as the implied relation between the form of the bed and my bed itself.

Similarly if I work on a market stall I might spend all day weighing and measuring and taking money and giving change. The numbers that I work with do not obviously have anything to do with the numbers the pure mathematician works with, and indeed she could say "my numbers are eternal and immutable and live in a mathematical heaven that I get at by pure thought, whereas your numbers are just the counters of your trade and have nothing to do with mine." Perhaps she is right, but if so, and if this also applies to the properties and qualities of sensible things—colour, weight, texture, species, genus and so on—what has knowledge, if knowledge can only be of such things, got to do with the world of everyday things we interact with? And who studies these forms; which branch of science?

Plato ends up with a dilemma. If belief and knowledge are of the same sorts of things, there does not seem to be any way of getting from the former to the latter. We cannot readily find a mental indicator or hallmark that allows us to say that any given belief is actually a piece of knowledge. But if they are of different sorts of things, then having knowledge (of the forms) does not seem to have anything to do with having beliefs and opinions (about what there is). There are three ways out of a dilemma. We can opt for one or the other horn of the dilemma, or we can try and steer a mid-path.

4.2 The mind or the senses?

Historically, opting for one or other horn has been most common. The first is to choose the mind and downplay the world of the senses, to argue that what is real and true is the world discovered by reason, by the eye of the mind. By contrast the world of the senses is a world of fleeting appearances and illusions, unworthy of the philosopher's attention. This certainly seems to be Plato's choice:

Socrates And the soul is like the eye; when resting upon that on which truth and being shine, the soul perceives and understands, and is radiant with intelligence; but when turned towards the twilight and to those things which come into being and perish, then she has opinion only, and goes blinking about, and is first of one opinion and then of another, and seems to have no intelligence. (*Republic*, p.508d)

This position is known as **rationalism**, because it privileges reason and rationality, the province of the mind, over the senses. Descartes, whose work we will look at in the next chapter, is counted as a rationalist.

Embracing the other horn of the dilemma leads to **empiricism**. Empiricists seek to account for our knowledge in terms of sense perception. Whereas rationalists tend to attribute powers to the mind to explain its capacity for knowledge, empiricists tend to regard the mind as a blank slate that is somehow written on in experience. The powers attributed to the mind by empiricists are much more limited, consisting in the main in capacities to compare and contrast perceptions and class them by similarities and differences. Empiricists deny that there are such things as forms, regarding them as intellectual illusions (and perhaps even delusions). Locke, Berkeley and Hume are all counted as empiricists. We will also see, later, a middle-way strategy proposed by Kant, whose work post-dates that of Descartes, Locke, Berkeley and Hume.

4.3 Sun, Cave, and Divided Line

The challenge for the rationalist is to explain not only what it is that we know (what the objects of knowledge are) but also how it is that we come to know them. When Socrates is pressed on this point in the *Republic*, he falls back on a series of analogies or illustrations, the Sun (p.507a–509c), the Divided Line (p.509d–511e), and the Cave (p.514a–517c). In each of these Plato tries to illustrate how the mind, unaided by the senses, can attain knowledge. He is at pains to point out that these are *analogies*, though, because he himself, in the guise of Socrates, lacks knowledge. In the *Meno*, as we saw, Plato put forward the doctrine that learning is recollection. This recurs in the Cave analogy, where Plato says:

Socrates But then, if I am right, certain professors of education must be wrong when they say that they can put knowledge into the soul that was not there before, like sight into blind eyes.

Glaucon They undoubtedly say this.

Socrates Whereas our argument shows that the power and capacity of learning exists in the soul already; and that just as if it were not possible to turn the eye from darkness to light without the whole body, so too the instrument of knowledge can only by the movement of the whole soul be turned from the world of becoming to that of being, and of the brightest

> and best of being, or in other words, of the good. (*Republic*, p.518b–d)

Exactly what Plato has in mind by "the good" is hard to tell. One possibility is this; our minds are themselves entities originally created by God, suggesting that there is a *form of the mind*, an ideal mind that is capable of attaining complete knowledge of all the **forms**. The characteristic or property of this form is a capacity to achieve knowledge of the forms. Insofar as our individual minds partake of this form we, too, are able to attain knowledge of the forms. So "the good" is the characteristic of the mind that makes it what it is, that enables us to attain knowledge and it is this more than anything else that we desire knowledge of.

In the Sun analogy, Plato says this:

> **Socrates** Sight being, as I conceive, in the eyes, and he who has eyes wanting to see; colour being also present in the objects, still unless there be a third nature specifically adapted to the purpose, sight, as you know, will see nothing and the colours will be invisible.
>
> **Glaucon** Of what nature are you speaking?
>
> **Socrates** Of that which you term "light". (*Republic*, p.507d–e)

Think of the form of the good, then, as a special sort of light that illuminates the world of the mind. When, with the mind's eye, we see things picked out by, or bathed in, this special sort of light, we have knowledge of these things. Of course this is an analogy because Plato is struggling to say what is pretty much beyond expression, but the idea is that the form of the good provides the mental indicator or hallmark that tells us when we have knowledge rather than opinion. So: what is the criterion of knowledge? When what I see with my mind's eye is bathed in the form of the good. The challenge for the rationalist is to make more of this than an analogy that the empiricist would most likely dismiss as a mere flight of fancy.

Chapter 5

Descartes' Rationalism

Nearly two thousand years separate Plato and Descartes, and it would be rather rash to skip two millenia without any comment at all. This period includes the other great Greek philosopher, Aristotle, along with sceptics, stoics, Epicureans, and a host of Christian and Moorish thinkers. In terms of theory of knowledge it would take us too far afield to discuss these figures, although Aristotle will crop up later in 6.2 and 8.1. So with a certain alacrity we will jump forward to the early seventeenth century. At this time the world we now inhabit was recognisably emerging through the work of Galileo and his contemporaries, an age in which, to over-simplify matters somewhat, it began to be accepted that knowledge could come from reading the book of nature as well as from reading the scriptures, the book of God.

Reading this book requires, as Galileo observes, a knowledge of the language it is written in:

> Philosophy is written in this grand book, the universe, which stands continually open to our gaze. But the book cannot be understood unless one first learns to comprehend the language and read the letters in which it is composed. It is written in the language of mathematics, and its characters are triangles, circles, and other geometrical figures without which it is humanly impossible to understand a single word of it; without these, one wanders about in a dark labyrinth. (quoted at Hall, *From Galileo to Newton*, p.84)

Put differently, qualitative Aristotelian natural science was displaced by a quantitative approach, underpinned by increasingly comprehensive and rigorous application of mathematical methods. This leads to the "disenchanting of nature", as attempts to come to terms with the world in terms of gods and spirits are displaced by differential equations and formalisms. One way of writing a history of the modern world would be to pinpoint increasingly sophisticated applications of mathematics (Newton's *Principia Mathematica* with its geometrical presentation of the methods of the calculus, Maxwell's use of differential equations in his work on the electromagnetic field in the nineteenth century, and in the twentieth century Einstein's use of tensor methods in relativity, and matrix methods in quantum mechanics) and correlate these with philosophical retreats. It is the changing role of philosophy relative to mathematics and physics after Galileo, with the emergence of a clear-cut distinction between philosophy and natural science, that is distinctive of the modern world. At least this is my excuse for somewhat glibly sliding over the two thousand-odd years that separate Plato and Descartes.

5.1 The scientific context

Descartes encourages his readers to approach his works with an open mind because, as he tells Mersenne in a letter, his intentions are less than transparent:

> I may tell you, between ourselves, that these six Meditations contain all the foundations of my physics. But please do not tell people, for that might make it harder for the supporters of Aristotle to approve them. I hope that readers will gradually get used to my principles, and recognise their truth, before they notice that they destroy the principles of Aristotle. (*Descartes' Correspondence*, AT III p.297–8)

The *Meditations*, published in 1641, is Descartes' most important work, and it is the one we will concentrate on here. It is a deceptive work, though, so my suggestion is, in a manner akin to the method of doubt Descartes proposes, to put to one side everything you have heard or read about Descartes and the *Meditations* and approach the text (or the extracts here) as if you're reading it for the very first time with no preconceptions, much as Descartes' contemporaries first read it in 1641. Although

Descartes' work contributed greatly to an emerging world-view that did indeed destroy Aristotle's principles, it still retains some recognisably Aristotelian ideas.

Nowadays institutional demarcations are much clearer and more rigid than they were in the seventeenth century. Physicists do physics, mathematicians do mathematics, and philosophers do whatever it is that philosophers do. In Descartes' time there were no such disciplinary boundaries so it is at best misleading to think of him solely as a philosopher writing philosophical works for philosophers. Rather he writes as an intellectual whose field of inquiry ranges across the human and the natural sciences, and addresses his work to an educated but equally generalist audience. Indeed he saw what we would regard as the distinctively philosophical as a fairly minor preliminary to the study of what now counts as the field of the natural sciences:

> I believe that it is very necessary to have properly understood, once-in-a-lifetime, the principles of metaphysics, since they are what gives us knowledge of God and the soul. But I think also that it would be very harmful to occupy one's intellect frequently in meditating upon them, since this would impede it from devoting itself to the functions of the imagination and the senses ...One should not devote so much effort to the *Meditations* and to metaphysical questions, or give them elaborate treatment in commentaries and the like. Still less should one do what some try to do, and dig more deeply into these questions than the author [Descartes himself] did; he has dealt with them quite deeply enough. It is sufficient to have grasped them once in a general way, and then to remember the conclusion. Otherwise, they draw the mind too far away from physical and observable things, and make it unfit to study them. Yet it is just these physical studies that it is most desirable for people to pursue, since they would yield abundant benefits for life. The author did follow up metaphysical questions fairly thoroughly in the *Meditations*, and established their certainty against the sceptics, and so on; so everyone does not have to tackle the job for himself, or need to spend time and trouble meditating on these things. (*Descartes' Correspondence*, AT III p.695, AT V p.165)

Metaphysics, literally "beyond physics", is the discipline in which certain principles are established and got out of the way before going on to do physics. These key principles are bound up with what we already know, what we can come to know, and how we get (or got) to know it. As we have seen, though, questions about knowledge are rarely cleanly separable from questions of what there is (the objects of knowledge). This applies as much to Descartes as to any other philosopher. What Descartes does is draw a boundary between metaphysics and physics, with God and the soul—loosely, the mind—taken as the subject of metaphysics. The world of material bodies (extended bodies, in Descartes' terminology, bodies that have physical extension and exist in space and time), is the subject of physics. This rigid separation of questions concerning God and the soul from questions concerning extended physical bodies is at the heart of Descartes' philosophy. This is why the treatment of metaphysical questions in the *Meditations* clears a path for Descartes' physics, which treats physical bodies as inert objects that, literally, bump into and push one another around:

> It was in a sense a bargain, attested by Descartes, that nature and the animals should be handed over to mechanistic science while God and the soul and all the ultimates rested with philosophers and divines. (Hall, *From Galileo to Newton*, p.340)

Descartes took this division entirely seriously, arguing that animals fall outside of metaphysics because they lack immortal souls. The essence of the *Meditations* is to prove a "real distinction" between mind (the human soul) and body (extended physical beings), and that the essence of the mind is to think, that of physical bodies to be extended in space and time (bodies in motion that are prone to collide with one another). But after separating them in this way Descartes found it extremely difficult to give a credible account of how minds and bodies interact with one another, as they so evidently do. Later philosophers argued either that there are only minds and their ideas (idealism), or that there are only material bodies and these somehow "realise" mindedness (materialism, behaviourism). But to pursue these issues would be to divert into the philosophy of mind, so we will stick with Descartes' dualistic approach. Consequently first there is the mind, and our knowledge of it (the subject of metaphysics), and second the outside world, and our knowledge of it (the subject of the natural sciences).

It is hard to over-emphasise the extent to which Descartes' thought marks the emergence of the modern world. This separation of mind and body is the "disenchantment of nature", with nature becoming a mechanical world of springs and levers and pulleys, of pushes and pulls between inert, mindless material bodies, treated quantitatively by mathematical physics. The world is no longer strange and mysterious, a world of elves and goblins and demons and things that go bump in the night. It is a world in which men come to grasp the forces that operate in order to bend nature to their will, with science as the means to bring this about, an approach that can be seen at work in Boyle's *Free Enquiry* (1686):

> As it more recommends the skill of an engineer to contrive an elaborate engine so as that there should need nothing to reach his ends in it but the contrivance of parts devoid of understanding, than if it were necessary that ever and anon a discreet servant should be employed to concur notably to the operations of this or that part, or to hinder the engine from being out of order; so it more sets off the wisdom of God in the fabric of the universe that he can make so vast a machine perform all those many things which he designed it should by the mere contrivance of brute matter, managed by certain laws of local motion and upheld by his ordinary and general concourse ...
>
> I consider the frame of the world as a great and, if I may so speak, pregnant automaton, that like a woman with twins in her womb, or a ship furnished with pumps, ordnance, etc. is such an engine as comprises or consists of several lesser engines. And this compounded machine, in conjunction with the laws of motion freely established and still maintained by God among its parts, I look upon as a complex principle, whence results the settled order or course of things corporeal. (*A Free Enquiry into the Vulgarly Received Notion of Nature*, p.11, 40)

5.2 God and the mind

To begin with the mind, Descartes initially claims that we are minded beings and that it makes no sense to deny this. If someone denies that they have a mind, should we take them seriously? Would we not think they are joking, or that they are mad, that they have literally lost their mind? Of course they might be an android in which case, appearances notwithstanding, what they say is indeed true.

Descartes argues that the essence of the mind, what is defining of it, is that it is a thing (a non spatio-temporal thing) that thinks. (Consequently the mind cannot be the brain, because the latter is a spatio-temporal thing.) The **essence** of something is what makes it what it is, such that if it loses its essence it ceases to be what it is. So he argues that the mind is always thinking, because if it ceased to think it would cease to be. What he means by "thought" is pretty wide ranging:

> It does not seem to me a fiction, but a truth which nobody should deny, that there is nothing entirely in our power except our thoughts; at least if you take the word "thought" as I do, to cover all the operations of the soul, so that not only meditations and acts of the will, but the activities of seeing and hearing and deciding on one movement rather than another, so far as they depend on the soul, are all thoughts. (*Descartes' Correspondence*, AT II p.36)

In this sense it is hard not to think that Descartes is right. What would it be to stop thinking? If you lose your train of thought you might say "my mind's a complete blank," but do you really mean this in the way that a computer screen goes blank when someone pulls the plug? Is a mind without a thought conceivable?

An immediate consequence is that our knowledge of our minds cannot be derived from sense-perception, because my mind is not the sort of thing that I can perceive. However I know it, or rather, however I know myself, has to be more immediate than this. If the scope of metaphysics is God and the soul—my soul—then whatever knowledge we have in metaphysics has to be in the mind, because there is no obvious way in which it can get in from the outside. (Talk of "outside" here is, of course, problematic, because minds are not spatio-temporal extended things, so there literally cannot be an "outside".) This leads Descartes to claim that whatever metaphysical truths we know about God and the soul must be **innate**, that is, they must have been in the mind from birth. But if these truths are innate, it does not follow immediately that we are aware of them (that we consciously know them):

> You [Voetius, professor of theology at Utrecht] claim that the arguments I use to prove the existence of God have force only for those who already know that he exists, since they depend simply on notions that are innate within us. But notice that all those things whose knowledge is said to be naturally implanted in us are not expressly known by us; they are merely such that we come to know them by the power of our own native intelligence, without any sensory experience. All geometrical truths are of this sort—not just the most obvious ones, but all the others, however abstruse they may appear. Hence, according to Plato, Socrates asks a slave-boy about the elements of geometry and thereby makes the boy able to dig out certain truths from his own mind which he had not previously recognised were there, thus attempting to establish the doctrine of reminiscence. Our knowledge of God is of this sort. (*Descartes' Correspondence*, AT VIIIB p.166–7)

Whatever we know about God and about the mind and its capacities has to be discovered by some sort of inner, mental exercise, and it is this knowledge that, if we follow the path laid out in the *Meditations*, Descartes thinks we will arrive at. Before turning to the *Meditations* themselves, though, three preliminary points need to be noted.

The first is the claim sometimes made by commentators on Descartes' behalf that the mind must be transparent to itself. What is in the mind, our knowledge of God and the other innate ideas we are born with, whatever they may be, must be available to the mind. But this is not what Descartes says. Indeed he points out that:

> Although the idea of God is imprinted on the human mind in such a way that everyone has within himself the power to know him, this does not prevent many people passing

> through their whole lives without ever having a distinct representation of this idea. (*Descartes' Correspondence*, AT IV p.187)

If the mind is transparent to itself than there would be no need for the *Meditations* because we would already have a grasp of what it is that Descartes seeks to help us uncover (or recollect). And indeed if the mind is so transparent there could hardly be disagreement about the nature of God, the mind, and the truths of metaphysics generally, and mistakes in these areas would be inexplicable.

Second, the scope of the *Meditations*. Its aim is to help us recollect the truths of metaphysics, not truths about the world of extended, physical bodies, beyond the claim that there is such a world. The *Meditations* is a preparation for experimental science, an activity that Descartes himself was heavily involved in.

Third, the demands the *Meditations* makes on its reader are intensely personal. These include a degree of psychological isolation because each individual mind is private and independent of any other mind (except possibly God's). My knowledge of what is innate in my mind is private, and I cannot learn anything about it from anybody else. This underlies the style and approach of the *Meditations*, which does not so much teach as assist the reader to recollect what is innate in their particular case. In this respect the *Meditations* is interestingly comparable to the *Meno*, because in both cases the authors have to know what it is that they are midwives to but they cannot teach what it is that they are trying to bring forth. Each of us has to come to see for ourselves.

Unfortunately misunderstandings of Descartes' views on these points, particularly the first, has led to a widespread caricature of his position. Descartes does not claim that the workings and contents of a mind are utterly transparent and accessible to the person concerned, that the mind is a "ghost in a machine", a numinous spirit that somehow permeates and operates a physical body. Further he sets strict limits on what is innate in the mind. It is only the mind in terms of its knowledge of metaphysics, doctrines concerning God and the intellectual aspects of the mind, that is the subject of the *Meditations*, and it is only this we will be considering.

5.3 The *Meditations*

First Meditation

Descartes begins the First Meditation, entitled "What can be called into doubt", thus:

> Several years have now elapsed since I first became aware that I had accepted, even from my youth, many false opinions for true, and that consequently what I afterward based on such principles was highly doubtful; and from that time I was convinced of the necessity of undertaking once in my life to rid myself of all the opinions I had adopted, and of commencing anew the work of building from the foundation, if I desired to establish a firm and abiding superstructure in the sciences ...
>
> But, to this end, it will not be necessary for me to show that the whole of these are false—a point, perhaps, which I shall never reach ...it will be sufficient to justify the rejection of the whole if I shall find in each some ground for doubt. Nor for this purpose will it be necessary even to deal with each belief individually, which would be truly an endless labour ...All that I have, up to this moment, accepted as possessed of the highest truth and certainty, I received either from or through the senses. I observed, however, that these sometimes misled us; and it is prudent not to place absolute confidence in that by which we have even once been deceived. (*Meditations*, AT VII p.17–8)

Clearly Descartes is engaged in a "one-off" ground-clearing exercise as a preliminary to his scientific views. By the end of this passage he has cast doubt on what is learnt via the senses, and this leads into a famous passage in which he suggests that the world as experienced is wholly an illusion and that he is in fact asleep the whole time. He then proposes the idea of an evil deceiver who maintains him permanently in this state:

> I will suppose, then, not that Deity, who is sovereignly good and the fountain of truth, but that some malignant demon, who is at once exceedingly potent and deceitful, has employed all his artifice to deceive me; I will suppose that

> the sky, the air, the earth, colours, figures, sounds, and all external things, are nothing better than the illusions of dreams, by means of which this demon has laid traps for my credulity; I will consider myself as without hands, eyes, flesh, blood, or any of the senses, and as falsely believing that I am possessed of these; I will continue resolutely fixed in this belief, and if by this means it is indeed not in my power to arrive at the knowledge of truth, I shall at least do what is in my power, viz., [suspend my judgment], and guard with settled purpose against giving my assent to what is false, and being imposed upon by this deceiver, whatever be his power and artifice. (*Meditations*, AT VII p.22–3)

Second Meditation

In this state Descartes ends the First Meditation and goes on to the Second, entitled "The nature of the human mind and how it is better known than the body." He picks up his previous thoughts thus:

> I assuredly existed, since I was persuaded. But there is I know not what being, who is possessed at once of the highest power and the deepest cunning, who is constantly employing all his ingenuity in deceiving me. Doubtless, then, I exist, since I am deceived; and, let him deceive me as he may, he can never bring it about that I am nothing, so long as I am conscious that I am something. So that it must, again, be maintained, all things being maturely and carefully considered, that this proposition *I am, I exist*, is necessarily true each time it is expressed by me, or conceived in my mind.
>
> But I do not yet know with sufficient clearness what I am, though my existence is assured; and hence, in the next place, I must take care, lest I carelessly substitute some other object in place of what is properly myself, and thus wander from truth, even in that knowledge which I hold to be of all others the most certain and evident [that I exist]. For this reason, I will now consider anew what I formerly believed myself to be, before I entered on the present train of thought; and of my previous opinion I will retreat from all that can in any way be invalidated by the grounds of doubt

I have adduced, in order that there may at length remain nothing but what is certain and indubitable.

What then did I formerly think I was? Undoubtedly I judged that I was a man. But what is a man? Shall I say a rational animal? Assuredly not; for it would be necessary forthwith to inquire into what is meant by animal, and what by rational, and thus, from a single question, I should insensibly glide into others, and these more difficult than the first; nor do I now possess enough leisure to warrant me in wasting my time amid subtleties of this sort. I prefer here to attend to the thoughts that sprung up of themselves in my mind, and were inspired by my own nature alone, when I applied myself to the consideration of what I was. In the first place, then, I thought that I possessed a face, hands, arms, and all the fabric of members that appears in a corpse, and which I called by the name of body. It further occurred to me that I was nourished, that I walked, perceived, and thought, and all those actions I referred to the soul; but what the soul itself was I either did not stay to consider, or, if I did, I imagined that it was something extremely rare and subtle, like wind, or flame, or ether, spread through my more solid parts.

As regards the body, I did not doubt of its nature, but thought I distinctly knew it, and if I had wished to describe it according to the notions I then entertained, I should have explained myself in this manner: by body I understand all that can be bounded by a certain figure; that can be comprised in a certain place, and so fill a certain space as therefrom to exclude every other body; that can be perceived either by touch, sight, hearing, taste, or smell; that can be moved in different ways, not indeed of itself, but by whatever else comes into contact with it; for the power of self-motion, as likewise that of perceiving and thinking, I held as by no means pertaining to the nature of body; on the contrary, I was somewhat astonished to find such faculties existing in certain kinds of bodies [i.e. human beings].

As to myself, what can I now say that I am, since I suppose there exists an extremely powerful and, so to speak, malignant being, whose endeavours are wholly directed toward

> deceiving me? Can I affirm that I possess any one of all those attributes of which I have lately spoken as belonging to the nature of body? After attentively considering them in my mind, I find none of them that can properly be said to belong to myself. To recount them were idle and tedious. Let us pass, then, to the attributes of the soul. The first mentioned were the powers of nutrition and walking; but, if it be true that I have no body, it is true likewise that I am capable neither of walking nor of being nourished. Perception is another attribute of the soul; but perception too is impossible without the body; besides, I have frequently, during sleep, believed that I perceived objects which I afterward observed I did not in reality perceive. Thinking is another attribute of the soul; and here I discover what properly belongs to myself. This alone is inseparable from me. I am—I exist: this is certain; but how often? As often as I think; for perhaps it would even happen, if I should wholly cease to think, that I should at the same time altogether cease to be. I now admit nothing that is not necessarily true. I am therefore, precisely speaking, only a thinking thing, that is, a mind, intelligence, understanding, or reason, terms whose signification was before unknown to me. I am, however, a real thing, and really existent ...
>
> But what, then, am I? A thinking thing, it has been said. But what is a thinking thing? It is a thing that doubts, understands, affirms, denies, wills, refuses; that imagines also, and perceives. (*Meditations*, AT VII p.25–7, 28)

Bearing in mind the scope of Descartes' notion of "thought" (5.2), in this passage he arrives at the essence of man—a thing that thinks, a mind, conceived of as entirely independent and separate of any physical (extended) body. In this way Descartes sets up his "real distinction" between mind and body, and he goes on to consider the nature of bodies in his well-known "block of wax" argument:

> Take, for example, this piece of wax; it is quite fresh, having been but recently taken from the beehive; it has not yet lost the sweetness of the honey it contained; it still retains somewhat of the odour of the flowers from which it was

gathered; its colour, figure, size, are apparent (to the sight); it is hard, cold, easily handled; and sounds when struck upon with the finger. Above all, the things that contribute to make a body as distinctly known as possible, are found in the one before us. But, while I am speaking, let it be placed near the fire–what remained of the taste exhales, the smell evaporates, the colour changes, its figure is destroyed, its size increases, it becomes liquid, it grows hot, it can hardly be handled, and, although struck upon, it emits no sound. Does the same wax still remain after this change? It must be admitted that it does remain; no one doubts it, or judges otherwise. What, then, was it I knew with so much distinctness in the piece of wax? Assuredly, it could be nothing of all that I observed by means of the senses, since all the things that fell under taste, smell, sight, touch, and hearing are changed, and yet the same wax remains.

It was perhaps what I now think, viz., that this wax was neither the sweetness of honey, the pleasant odour of flowers, the whiteness, the figure, nor the sound, but only a body that a little before appeared to me conspicuous under these forms, and which is now perceived under others. But, to speak precisely, what is it that I imagine when I think of it in this way? Let it be attentively considered, and, taking away everything that does not belong to the wax, let us see what remains. There certainly remains nothing, except something extended, flexible, and movable. But what is meant by flexible and movable? Is it not that I imagine that the piece of wax, being round, is capable of becoming square, or of passing from a square into a triangular figure? Assuredly such is not the case, because I conceive that it admits of an infinity of similar changes; and I am, moreover, unable to compass this infinity by imagination, and consequently this conception which I have of the wax is not the product of the faculty of imagination. But what now is this extension? Is it not also unknown? For it becomes greater when the wax is melted, greater when it is boiled, and greater still when the heat increases; and I should not conceive [clearly and] according to truth, the wax as it is, if I did not suppose that the piece we are considering admitted even of a wider variety of extension than I ever imagined,

> I must, therefore, admit that I cannot even comprehend by imagination what the piece of wax is, and that it is the mind alone which perceives it. (*Meditations*, AT VII p.30–1)

This is curious, and it might well be thought that Descartes labours the point because he has an ulterior motive. His idea is that stripping one's idea of a physical body of all the qualities that arise from perception (smell, taste, colour, texture, and so on) leaves just the idea of a quantity of matter that occupies space. Descartes regards this property of bodies, their extension in space, as the **essence** of bodies, and this makes them categorically different from minds, which are not extended bodies and are not obviously located in space. In this way Descartes allocates the mind to metaphysics and physical bodies to physics. Physical bodies are essentially inert occupiers of space, and this leads directly to a physics based on the mechanical interactions (motions and impacts) of extended bodies.

Third Meditation

In the Third Meditation, titled "The existence of God", Descartes sets out the first of his two proofs of God (the second is in the Fifth Meditation). It is the role Descartes assigns to God rather than the proofs themselves that is of interest, so they will only be touched on briefly. Descartes faces two difficulties; first, to explain the origins of the contents of the mind that the *Meditations* proposes to uncover, and second, to show that there is a world external to the mind that our mental contents are related to. For the first Descartes proposes a theory of innate ideas, including an idea of God that God himself has planted in us. The second is bound up with the first, because Descartes claims that God is an omniscient benevolent being who would not have implanted false ideas in us. It follows that correctly grasping our innate ideas is a prelude to making true judgments about the world around us. The role assigned to God is of explaining how it is that we know, or at least seem to know, more than we can ever derive from the senses alone. Faced with much the same question Socrates falls back (in the *Meno*) on a "glorious truth" of reincarnation and recollection (2.1). The truth is that there is something that needs to be explained, a set of principles that we seem to have in mind before we can make sense of what we get in perception, and which we cannot get out of perception. Although they worshipped different gods there are striking similarities between Descartes' and Socrates' approaches here.

The notion of "**innate ideas**" is introduced early in the Third Meditation as a hypothesis:

> But among these ideas, some appear to me to be innate, others adventitious, and others to have been invented by myself (factitious); for, as I have the power of conceiving what is called a thing, or a truth, or a thought, it seems to me that I hold this power from no other source than my own nature; but if I now hear a noise, if I see the sun, or if I feel heat, I have all along judged that these sensations proceeded from certain objects existing out of myself; and, in truth, it appears to me that sirens, hippogryphs, and the like, are inventions of my own mind. But I may even perhaps come to be of the opinion that all my ideas are of the class which I call adventitious, or that they are all innate, or that they are all factitious; for I have not yet clearly discovered their true origin. (*Meditations*, AT VII p.37–8)

Descartes goes on to consider his idea of God, proposing a purely conceptual argument that a cause must be more powerful than its effect. Since his idea of God is of an all-powerful, all-seeing being, perfect in every respect, it follows that God himself is at least all this, and most likely inconceivably more extraordinary:

> It is absolutely necessary to conclude from [the fact of my existence] alone that I am, and possess the idea of a being absolutely perfect, that is, of God, that his existence is most clearly demonstrated. There remains only the inquiry as to the way in which I received this idea from God; for I have not drawn it from the senses, nor is it even presented to me unexpectedly, as is usual with the ideas of sensible objects, when these are presented or appear to be presented to the external organs of the senses; it is not even a pure production or fiction of my mind, for it is not in my power to take from or add to it; and consequently there but remains the alternative that it is innate, in the same way as is the idea of myself. And, in truth, it is not to be wondered at that God, at my creation, implanted this idea in me, that it might

> serve, as it were, for the mark of the workman impressed on his work. (*Meditations*, AT VII p.51)

Fourth Meditation

It does not matter too much whether or not you accept the argument, although it does have a certain elegance. What is of interest is the capacities that we seem to have that Descartes is trying to account for, capacities he discusses further in the Fourth Meditation, titled "Truth and falsity." Descartes puts forward the idea that whatever he has "clearly and distinctly" before the mind is true, where to say that I have something "clearly and distinctly" before the mind is comparable to saying that it is bathed in the light of the good (as 4.3). This notion serves for Descartes as a mark of truth:

> It is certain, however, that we will never mistake the false for the true provided we give our assent only to what we clearly and distinctly perceive ...A perception which can serve as the basis for a certain and indubitable judgment needs to be not merely clear but also distinct. I call a perception "clear" when it is present and accessible to the attentive mind ...I call a perception "distinct" if, as well as being clear, it is so sharply separated from all other perceptions that it contains within itself only what is clear. (*Principles of Philosophy*, §§43, 45)

This is Descartes' criterion of truth which, for the moment, we will accept and see where it takes us. Descartes begins the Fourth Meditation with these words:

> I have been habituated these past few days to detach my mind from the senses, and I have accurately observed that there is exceedingly little which is known with certainty concerning corporeal objects, that we know much more of the human mind, and still more of God himself. I am thus able now without difficulty to abstract my mind from the contemplation of [sensible or] imaginable objects, and apply it to those which, as disengaged from all matter, are purely intelligible. And certainly the idea I have of the human mind in so far as it is a thinking thing, and not extended in length,

> breadth, and depth, and participating in none of the properties of body, is incomparably more distinct than the idea of any corporeal object. When I consider that I doubt, in other words, that I am an incomplete and dependent being, the idea of a complete and independent being, that is to say of God, occurs to my mind with so much clearness and distinctness, and from the fact alone that this idea is found in me, or that I who possess it exist, the conclusions that God exists, and that my own existence, each moment of its continuance, is absolutely dependent upon him, are so manifest, as to lead me to believe it impossible that the human mind can know anything with more clearness and certitude. And now I seem to discover a path that will conduct us from the contemplation of the true God, in whom is contained all the treasures of science and wisdom, to the knowledge of the other things in the universe.
>
> For, in the first place, I discover that it is impossible for him ever to deceive me, for in all fraud and deceit there is a certain imperfection: and although it may seem that the ability to deceive is a mark of subtlety or power, yet the will testifies without doubt of malice and weakness; and such, accordingly, cannot be found in God.
>
> In the next place, I am conscious that I possess a certain faculty of judging [or discerning truth from error], which I doubtless received from God, along with whatever else is mine; and since it is impossible that he should will to deceive me, it is likewise certain that he has not given me a faculty that will ever lead me into error, provided I use it aright.
>
> There would be no further doubt on this head, did it not seem to follow from this, that I am incapable of ever going wrong. (*Meditations*, AT VII p.52–4)

Descartes ends up with a key difficulty for rationalism, which we will call the **omniscience problem**. If it is all in the mind, and given that what is in the mind is true—it cannot very well be false, after all, because there is nothing that can make it false—how can error, falsehood, arise? Descartes' answer is based on a disparity between the intellect and the will:

> Whereupon, regarding myself more closely, and considering

> what my errors are (which alone testify to the existence of imperfection in me), I observe that these depend on the concurrence of two causes, viz., the faculty of cognition, which I possess, and that of election or the power of free choice,—in other words, the understanding and the will. For by the understanding alone, I [neither affirm nor deny anything but] merely apprehend the ideas regarding which I may form a judgment; nor is any error, properly so called, found in [my understanding] thus accurately taken. And although there are perhaps innumerable objects in the world of which I have no idea in my understanding, it cannot, on that account be said that I am deprived of those ideas [as of something that is due to my nature], but simply that I do not possess them, because, in truth, there is no ground to prove that God ought to have endowed me with a larger faculty of cognition than he has actually bestowed upon me ...Nor, moreover, can I complain that God has not given me freedom of choice, or a will sufficiently ample and perfect, since, in truth, I am conscious of will so ample and extended as to be superior to all limits. And what appears to me here to be highly remarkable is that, of all the other properties I possess, there is none so great and perfect as that I do not clearly discern it could be still greater and more perfect ...
>
> From all this I discover, however, that neither the power of willing, which I have received from God, is of itself the source of my errors, for it is exceedingly ample and perfect in its kind; nor even the power of understanding, for as I conceive no object unless by means of the faculty that God bestowed upon me, all that I conceive is doubtless rightly conceived by me, and it is impossible for me to be deceived in it. Whence, then, spring my errors? They arise from this cause alone, that I do not restrain the will, which is of much wider range than the understanding, within the same limits, but extend it even to things I do not understand, and as the will is of itself indifferent to such, it readily falls into error and sin by choosing the false in place of the true, and evil instead of good. (*Meditations*, AT VII p.56–7, 58)

When I go wrong it is because I make judgments where I lack clear and distinct ideas, so in principle I need only make sure that I have such ideas and all will be well. But in fact in the paragraph before the last quote from Descartes I deliberately slid over a problem that you probably noticed—if what is in the mind, our innate ideas, cannot be false because there is not anything to make them false, then equally it cannot be said that they are true, because there is not anything to make them true. Consequently innate ideas cannot be true or false, they can only be more or less clearly and distinctly perceived (by the mind). Erroneous judgements—falsehoods—arise only when I deploy unclear and indistinct ideas in thinking and talking about the world, because I have wilfully failed to get the relevant ideas clear and distinct in my mind first.

We have seen Descartes make a distinction between the mind as a thinking thing, with a stock of innate ideas, and bodies as inert chunks of extended matter that occupy space. The "block of wax" argument is intended to delimit what we can know of extended bodies solely on the basis of what is in the mind. It follows from this argument (Second Meditation, above) that it is *only* innate ideas that can be clearly and distinctly perceived. Because these ideas are in the mind and are entirely independent of our experience of the world, if we can devise a suitable method we should be able to get to know them entirely and completely. Descartes' method of doubt is designed with just this aim in mind.

Fifth Meditation

Descartes discusses the extent of **innate ideas**, early in the the Fifth Meditation, titled "The essence of material things, and the existence of God considered the second time":

> But before considering whether such objects as I conceive exist without me, I must examine their ideas in so far as these are to be found in my consciousness, and discover which of them are distinct and which confused.
>
> In the first place, I distinctly imagine that quantity which the philosophers commonly call continuous, or the extension in length, breadth, and depth that is in this quantity, or rather in the object to which it is attributed. Further, I can enumerate in it many diverse parts, and attribute to each of these all sorts of sizes, figures, situations, and local motions;

> and I can also assign to each of these motions all degrees of duration.
>
> And I not only distinctly know these things when I thus consider them in general; but besides, by a little attention, I discover innumerable particulars respecting figures, numbers, motion, and the like, which are so evidently true, and so accordant with my nature, that when I now discover them I do not so much appear to learn anything new, as to call to remembrance what I before knew, or for the first time to remark what was before in my mind, but to which I had not hitherto directed my attention.
>
> And what I here find of most importance is, that I discover in my mind innumerable ideas of certain objects, which perhaps possess no reality beyond my thought but nevertheless cannot be considered to be nothing. Although these ideas are not invented by me, and whether or not it is in my power to think or not to think them, they still possess true and immutable natures of their own. As, for example, when I imagine a triangle, although there is not perhaps and never was in any place in the universe apart from my thought one such figure, it remains true nevertheless that this figure possesses a certain determinate nature, **form**, or **essence**, which is immutable and eternal, and not invented by me, nor in any degree dependent on my thought.
>
> Nor is it a valid objection to allege, that perhaps this idea of a triangle came into my mind by the medium of the senses, through my having seen bodies of a triangular figure; for I am able to form in thought an innumerable variety of figures with regard to which it cannot be supposed that they were ever objects of sense ...I recollect that even when I still strongly adhered to the objects of sense, I reckoned among the number of the most certain truths those I clearly conceived relating to figures, numbers, and other matters that pertain to arithmetic and geometry, and in general to pure and abstract mathematics. (*Meditations*, AT VII p.63–5)

He goes on to offer a further proof of God which we can safely skip. But the above passage is important because Descartes claims that both pure *and* applied mathematics are innate. His physics is based on extended bodies that occupy and move relative to one another in space, and collide with one another. Mathematics, particularly geometry, provides a framework for this conception of physics; the mathematical principles of geometry with one or two extensions (principally the notion of force) become the laws of nature for a mechanical world. As we saw in the Second Meditation, it is of the **essence** of material bodies that they are extended in space, that they move relative to, and collide with, one another. In this way Descartes sets the scene for modern science with its emphasis on applying mathematical methods to physical bodies that are inert, that is, do not think and do not exercise free will. In Descartes' book only humans have free will, all else is subject to (mathematical) laws.

Sixth Meditation

But establishing the essence of material bodies in terms of our ideas of them does not show that there are any such bodies outside of us. This is the topic of the Sixth (and final) Meditation, titled "The existence of material things, and the real distinction between mind and body." The opening paragraph is this:

> There now only remains the inquiry as to whether material things exist. With regard to this question, I at least know with certainty that such things may exist, in so far as they constitute the object of pure mathematics, since, regarding them in this aspect, I can conceive them clearly and distinctly. For there can be no doubt that God possesses the power of producing all the objects I am able distinctly to conceive, and I never considered anything impossible to him, unless when I experienced a contradiction in the attempt to conceive it aright. Further, the faculty of imagination which I possess, and of which I am conscious that I make use when I apply myself to the consideration of material things, is sufficient to persuade me of their existence: for, when I attentively consider what imagination is, I find that it is simply a certain application of the cognitive faculty to a body which is immediately present to it, and which therefore exists. (*Meditations*, AT VII p.71–2)

So far, what we know innately about material bodies (that they are extended in space, can move relative to and collide with one another, and that we can use applied mathematics to theorise about them) has come purely from reflection on what we have in mind. Consequently it is still possible that there is no physical world at all outside of my mind. The Harvard philosopher Hilary Putnam proposed as a thought-experiment that perhaps I am just a **brain in a vat**, the subject of a scientific experiment in which what I take to be the outside world is just a product of electrical stimulations applied to my brain by a mad scientist—Descartes' all-deceiving evil demon brought to life. How can I possibly tell that it is not so? Of course if he does the job properly there is no way I can tell. Descartes ultimately counters the "evil demon" argument with the claim that God is no deceiver, that the world I perceive is truly "out there":

> But I am accustomed to imagine many other objects besides that corporeal nature which is the object of pure mathematics as, for example, colours, sounds, tastes, pain, and the like, although with less distinctness; and, inasmuch as I perceive these objects much better by the senses, through the medium of which and of memory, they seem to have reached the imagination, I believe that, in order the more advantageously to examine them, it is proper I should at the same time examine what sense-perception is, and inquire whether from those ideas that are apprehended by this mode of thinking (consciousness), I cannot obtain a certain proof of the existence of corporeal objects. (*Meditations*, AT VII p.74)

He goes on to say:

> I cannot doubt but that there is in me a certain passive faculty of perception, that is, of receiving and taking knowledge of the ideas of sensible things; but this would be useless to me, if there did not also exist in me, or in some other thing, another active faculty capable of forming and producing those ideas. But this active faculty cannot be in me [in as far as I am but a thinking thing], seeing that it does not presuppose thought, and also that those ideas are frequently produced in my mind without my contributing to it in any way, and even frequently contrary to my will. This faculty must therefore exist in some substance different from me, in which all the objective reality of the ideas

that are produced by this faculty is contained . . .this substance is either a body, that is to say, a corporeal nature in which is contained . . .or it is God himself, or some other creature, of a rank superior to body . . .But as God is no deceiver, it is manifest that he does not of himself and immediately communicate these ideas to me, nor even by the intervention of any creature in which their objective reality is not . . .contained. For as he has given me no faculty whereby I can discover this to be the case but, on the contrary, a very strong inclination to believe that those ideas arise from corporeal objects, I do not see how he could be vindicated from the charge of deceit, if in truth they proceeded from any other source, or were produced by other causes than corporeal things: and accordingly it must be concluded, that corporeal objects exist. They may not be perhaps exactly such as we perceive by the senses, for their comprehension by the senses is, in many instances, very obscure and confused; but it is at least necessary to admit that all which I clearly and distinctly conceive as in them, that is, generally speaking all that is comprehended of them in the terms of pure mathematics, really exists external to me.

But with respect to other things which are either only particular, as, for example, that the sun is of such a size and figure, etc., or are conceived with less clearness and distinctness, as light, sound, pain, and the like, although they are highly dubious and uncertain, nevertheless on the ground alone that God is no deceiver, and that consequently he has permitted no falsity in my opinions which he has not likewise given me a faculty of correcting, I think I may with safety conclude that I possess in myself the means of arriving at the truth. In the first place, it cannot be doubted that in each of the dictates of nature there is some truth: for by nature, considered in general, I now understand nothing more than God himself, or the order and disposition established by God in created things; and by my nature in particular I understand the assemblage of all that God has given me.

But there is nothing that nature teaches me more expressly [or more sensibly] than that I have a body . . .Nature likewise teaches me by these sensations of pain, hunger, thirst, etc., that I am not merely lodged in my body as a pilot in a

> vessel, but that I am besides so intimately conjoined, and as it were intermixed with it, that my mind and body compose a certain unity. For if this were not the case, I should not feel pain when my body is hurt, seeing I am merely a thinking thing, but should perceive the wound by the understanding alone, just as a pilot perceives by sight when any part of his vessel is damaged; and when my body has need of food or drink, I should have a clear knowledge of this, and not be made aware of it by the confused sensations of hunger and thirst: for, in truth, all these sensations of hunger, thirst, pain, etc., are nothing more than certain confused modes of thinking, arising from the union and apparent fusion of mind and body. (*Meditations*, AT VII p.79–81)

This marks, more or less, the end of the *Meditations*. Descartes thinks that he has established the nature of, and our knowledge of, God and our own minds, and our knowledge of the aspects of the physical world (of extended bodies) that serve as a basis for a purely mechanical physics. His proofs of God serve to show that there is an external world that acts upon me and that my mind is somehow joined to my body. He establishes a demarcation of metaphysics and physics, allocating God and the mind, and physical bodies conceived of in terms of innate ideas, to metaphysics. Knowledge of what there is in terms of sense-perception and reflection on sense-perception is allocated to physics, to experimental science.

But if my body is just another extended body in space that bumps into other bodies, how is it that I perceive other bodies as having colour, texture, smell and so on? This leads to a famous philosophical distinction, between **primary** and **secondary qualities**. Descartes draws this distinction in Part One of his *Principles of Philosophy*:

> In order to distinguish what is clear in this connection from what is obscure, we must be very careful to note that pain and colour and so on are clearly and distinctly perceived when they are regarded merely as sensations or thoughts. But when they are judged to be real things existing outside our mind, there is no way of understanding what sort of things they are ...This will be especially clear if we consider the wide gap between our knowledge of those features

> of bodies which we clearly perceive ...and our knowledge of those features which must be referred to the senses. To the former class belong the size of the bodies we see, their shape, motion, position, duration, number and so on ...To the latter class belong the colour in a body, as well as pain, smell, taste and so on.
>
> It is true that when we see a body we are just as certain of its existence in virtue of its having a visible colour as we are in virtue of its having a visible shape; but our knowledge of what it is for the body to have a shape is much clearer than our knowledge of what it is for it to be coloured ...It is clear, then, that when we say that we perceive colours in objects, this is really just the same as saying that we perceive something in the objects whose nature we do not know, but which produces in us a certain very clear and vivid sensation which we call the sensation of colour. (*Principles of Philosophy*, §§68–70)

The **primary qualities** of a body go together with our innate ideas of it, whereas the **secondary qualities** go together with our sensory awareness of it. In a manner of speaking the world as it is "in itself" is a world of colourless lumps of matter moving around, and all its sensual richness is given to it by us in the act of sensing it.

It follows, on Descartes' approach, that in thinking about what is in my own mind, my innate ideas, I can spin out the whole of mechanics, the part of physics that deals with extended bodies in motion and in collision with one another, without needing to look at the world around me. Chemistry, biology and medicine are still experimental sciences, but physics somehow is not. This may seem odd but in fact a great deal of physics is and has always been done by thought experiment and later confirmed (or disconfirmed) by physical experiment. In truth there is no clear-cut distinction between physics and philosophy. Einstein's theory of relativity, for example, started off as a thought experiment and was only later confirmed experimentally (the phenomena of deflection of light due to gravitational fields, observed by Eddington *et al.* in 1919, and the "red shift" effect, connected with the propagation of light from distant heavenly bodies, observed by Adams in 1924; c.f. Einstein's *Relativity*, Appendix 3). In terms of knowledge, this question—of what we can get out of our own heads, and what has to come in from outside—became of central importance in philosophy after Descartes. And because philosophy was not then distinct from natural science, it was effectively a central question for science too.

5.4 Descartes and scepticism

The reading of Descartes I have given so far is somewhat unusual, because Descartes' philosophy is more usually presented as a response to scepticism. On this approach the *Meditations* is read as first proposing a particularly *philosophical* form of doubt. Philosophical doubt is systematic, and is concerned with particular sorts of questions. In everyday contexts you might doubt, for example, that the weather will stay fine for the match this afternoon or that your horse will win the 3.30 at Epsom. You may have reasons for your doubts in such cases, but these are not the sorts of doubts philosophers are concerned with. Doubting in philosophy is concerned rather with systematic questions, with whether we can be systematically in error, with whether it is possible that we are systematically deceived in perception or whether it is possible that what appears to be my waking life is in fact one long dream.

The difference between the reading of Descartes I have presented here—the "defence of Cartesian physics" reading—and the more usual "response to scepticism" reading lies more in the emphasis than in the detail. I would argue that the "defence of Cartesian physics" reading sets up a basis for telling a compelling story about the development of philosophy and science in the modern period, whereas the "response to scepticism" reading leaves Descartes largely marooned in a philosophical backwater. The development of science since Descartes' time is, after all, significantly more important than scepticism. Nevertheless the "response to scepticism" reading is well established and, at the least, one has to gloss over it.

Roughly, the schema of the "response to scepticism" reading is this. In the First Meditation Descartes pushes doubt, ordinary doubt first, then philosophical doubt, to a point at which all seems uncertain; maybe I am a **brain in a vat**, maybe what I think of as my waking life is one long dream (the "evil deceiver" argument). This is famously known as the **dreaming argument**. Descartes draws on the fact that we are sometimes deceived by our own senses, that a stick in water looks bent, for example, or a round tower seen from a distance looks square (the "argument from illusion/deception"). This undermines any argument that sensory knowledge is basic and beyond criticism, so Descartes finds he has to look elsewhere for some point of certainty.

In the Second Meditation Descartes establishes an indubitable truth, "I think, therefore I am"—I am a thing that thinks, so I must exist while I am thinking. This is a product of pure thought and is immune to any sort of doubt whatsoever, so it can serve as a fixed point from which we can, by some suitable process, establish further truths about ourselves and about what there is. Descartes establishes as a criterion of truth that a truth is whatever is "clearly and distinctly" before my mind, as this is the characteristic he identifies in the so-called **cogito**; *cogito, ergo sum*, I think, therefore I am. The rest of the *Meditations* is then read as a progressive response to sceptical challenges: our knowledge of the possibility of physical bodies is established ("block of wax" argument), God's existence is proved, and that he is no deceiver, hence there is an external world. As to the union of mind and body, this remains an insuperable problem that Descartes worked on for the rest of his life.

I would suggest now that you read back over the passages quoted from Descartes' works in the previous section, ignoring my commentary, with the "response to scepticism" approach in mind. How do the passages quoted read to you now? Of course different readings and interpretations of texts are grist to the philosopher's mill, and all that is needed is a good argument. Unsurprisingly I think the "defence of Cartesian physics" reading is more defensible than the "response to scepticism" approach, as I hope will become clear in the rest of this book.

Here is some more food for thought. A central theme in epistemology is the ways in which the beliefs we hold colour our "take" on new situations; whether there is a "right" way of perceiving something or other, or whether our "take" on whatever-it-is is inevitably coloured by our stock of existing beliefs. On a second reading of these passages, with the "response to scepticism" reading in mind, how did your approach to Descartes' arguments change? Assuming that it did change—I'd be very surprised if it did not—you could think of this as a case-study of the ways in which our existing beliefs affect our responses to new situations, that perhaps there is no "absolute" or "true" take on whatever-it-is.

So much for Descartes. There are lots of questions that I hope will still be bothering you. For example, what of innate ideas? If there are such things, where did they come from and how do we get to know them? How extensive are they? You might find Descartes' invocation of God troubling; but how else can we account for our knowledge of mathematics? Can we learn mathematics via the senses? Must we have knowledge of some sort prior to experience, that we bring to bear in experience, in order to make sense of our experience? Is the world really made up of colourless lumps of matter moving relative to and colliding with one another, with colour, smell, taste and so on superimposed by us, a **veil of perception** through which we perceive an inherently bleak and colourless world?

At this point historical introductions to theory of knowledge usually move on to Locke and empiricism. I think this approach is rather narrow, because it assumes that theory of knowledge is a philosophical preserve that can be mapped irrespective of what knowledge is of, what the point of the subject is. We seek knowledge of the world around us so that we can understand and co-operate with it in achieving whatever it is we seek to achieve. Traditionally philosophy divided into moral philosophy and natural philosophy, where the latter includes metaphysics, epistemology (theory of knowledge) and what we would now think of as the natural sciences. There is no clear distinction made by Plato and Descartes, or by Locke, Berkeley, Hume and Kant, between metaphysics, epistemology, and science, and my own view is that a study of Newton is helpful because his work profoundly influenced the philosophers who came after him. Such a study should also help to cement the narrative put forward here of a continuity between philosophy and physics. In Locke's "Epistle to the Reader" in his *Essay Concerning Human Understanding*, published two years after Newton's *Principia Mathematica*, Locke says:

> The Commonwealth of Learning, is not at this time without master builders, whose mighty designs, in advancing the sciences, will leave lasting monuments to the Admiration of posterity; But everyone must not hope to be a Boyle, or a Sydenham; and in an age that produces such masters,

> as the great Huygens, and the incomparable Mr Newton, with some other of that strain; it is ambition enough to be employed as an under-labourer in clearing ground a little, and removing some of the rubbish, that lies in the way to knowledge. (*Essay*, p.9–10)

There are two reasons why Newton profoundly influenced later philosophers. The first is that Newton's mathematical sophistication was (and for most of us remains) beyond the scope of his contemporaries' mathematical capabilities. This is reflected in Locke's remarks (Locke was one of Newton's correspondents and read the *Principia*, but had to ask Huygens to vouch for the mathematics as he could not follow it). Newton permanently redrew the map of speculative possibilities, and nobody after Newton could write philosophy in the Cartesian style. The second reason is that Newton's work, as we will see in the next chapter, focused attention on what is innate and what is derived from experience, thereby forcing a more sophisticated philosophical treatment of these topics. Before moving on to Newton, though, there is a piece of technical terminology it is useful to become familiar with, the distinction between *a priori* and *a posteriori* knowledge.

5.5 *A priori* and *a posteriori*

A priori knowledge is knowledge that is prior to experience, the sort of knowledge that Plato thinks we arrive at by recollection, that Descartes thinks of as **innate**. It is not knowledge that is learnt from experience, or is straightforwardly taught or demonstrated. It is true that we may, and in general do, come to realise that we possess such knowledge because it is shown to us or we are somehow instructed in it, by seeing a proof in geometry or being talked through a proof as Socrates talks the slave-boy through the example of doubling the area of the square in the *Meno*. The idea is that what we come to realise goes beyond what we have been shown; as I said in the Introduction, perhaps the key point in theory of knowledge is to explain how it is that we seem to know far more than we can straightforwardly explain in terms of experience alone. The slave-boy picks up a geometrical principle that is widely applicable, he learns a technique that he can apply in many different sets of circumstances.

What is proven also seems to possess a quality of **necessity**, characterised by the fact that there does not seem to be any way in which things could be otherwise. If you think of all plane right-angled triangles, is it possible that one of them does not have a long side (hypotenuse) that is not the square root of the sum of the squares of the two short sides? What would you think if you came across a right-angled triangle for which this is not the case? That some mistake has been made, or that somehow the principles of geometry have been suspended in this particular location in space and time? What would count as a rational explanation of such a phenomenon?

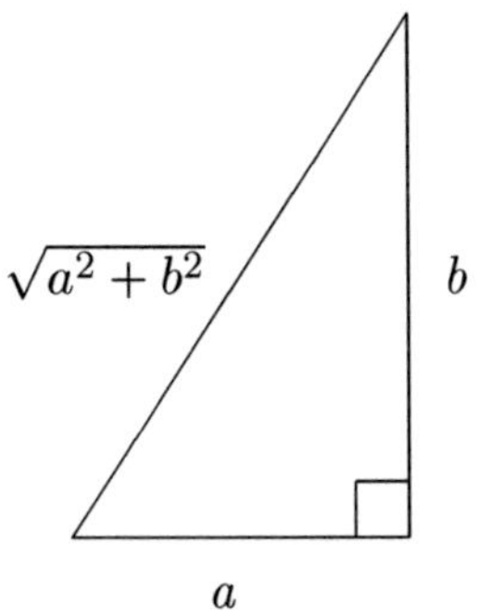

Pythagoras' Theorem

If there is *a priori* knowledge it does not follow without some further argument that it is innate, but if it is not innate, either some other account must be given of how we come by it or it has to be denied that there is *a priori* knowledge. Any one of these three options is possible but as always in philosophy whichever one is adopted (*a priori* and innate, *a priori* but not innate (acquired by some other means), no *a priori* knowledge) has to be argued for.

A posteriori knowledge is simply knowledge that is not *a priori*. In other words *a posteriori* knowledge is knowledge drawn from experience in such a way that it is explicable in terms of experience alone, that there is no "extra ingredient" purportedly derivable only from the mind. *A posteriori* knowledge lacks the necessary character of the *a priori* because it is always conceivable that our experience could have been otherwise. Is it inconceivable that the sun will not rise tomorrow, or that the table in front of you will not turn into a marshmallow in five seconds' time? Just because neither of these events has happened before does not mean that they will not (or cannot) happen in the future, so such events are **contingent**. As we will see when we turn to Locke's empiricism the disagreements between **rationalism** and **empiricism** principally involve differing views on the *a priori* and the *a posteriori*, and this is why Newton is useful as a way of setting up the argument.

Chapter 6

Newton; Rationalist or Empiricist?

6.1 Laws of nature

The title of Newton's great work, the *Principia Mathematica, The Mathematical Principles of Natural Philosophy*, is a deliberate allusion to Descartes' *Principia Philosophiae* (*Principles of Philosophy*). Newton's book was published in 1687, 43 years after Descartes', but Newton had studied both this work and Descartes' earlier *Geometry* carefully.

It was suggested in the Introduction that the central difficulty in theory of knowledge is that we seem to know more than we can account for on the basis of sensory evidence alone. One aspect of this is the regular, predictable behaviour of physical bodies; the sun will rise every morning, this chair will support my weight, the light will go on when I flick the switch, and so on. In such cases it certainly seems that physical laws are obeyed. Indeed it even seems odd to talk of laws being "obeyed" because there is no clear sense in which a chair or an electric circuit can disobey in the way that I can choose to disobey a direct order. We tend to take the regularities that physical bodies exhibit for granted and only notice when the unexpected happens—when something goes awry.

The title of Newton's book suggests that it is specifically *mathematical* principles he is interested in, and that his interest is limited to natural science, that is, physical science (as opposed to what used to be called "moral science", meaning the human and social sciences as well as philosophy as we now think of it). The regularities exhibited by physical bodies are usually theorised about mathematically; immensely complex mathematical models are used to design the shape of an aerofoil or the flow of the gases through the cylinder head of a car engine, for example. Newton concentrates mainly on the motions of bodies, particularly the heavenly bodies, and he begins his work with a series of eight definitions followed by three "Axioms, or the laws of motion":

Law 1 Every body perseveres in its state of rest, or of uniform motion in a [straight] line, unless it is compelled to change that state by forces impressed thereon.

Law 2 The alteration of motion is ever proportional to the motive force impressed; and is made in the direction of the [straight] line in which that force is impressed.

Law 3 To every action there is always opposed an equal reaction: or the mutual actions of two bodies upon each other are always equal, and directed to contrary parts. (*Principia Mathematica*, p.416–7/p.70–1)

These laws have an air of being self-evident; once they been stated, it is tempting to say that they are obvious. But now consider this—are they, in the sense set out at the end of the last chapter, *a priori* or *a posteriori*? If they are *a posteriori* then there must be some experience they can be derived from. Here is Newton's own remark on his First Law:

> Projectiles persevere in their motions, so far as they are not retarded by the resistance of the air, or impelled downwards by the force of gravity. A [spinning] top . . . does not cease its rotation, otherwise than as it is retarded by the air. The greater bodies of the planets and comets, meeting with less resistance in more free spaces, preserve their motions both progressive and circular for a much longer time. (*Principia Mathematica*, p.416/p.71)

In short, the law is hedged around with qualifications because no-one has ever, or indeed will ever, see a body in a state of rest or uniform straight-line motion that is not acted on by some external force. A body at rest on or near the surface of the earth is acted on by whatever supports it (stops it from falling). Even in deep space a projectile is subject to gravitational fields from distant stars. Rather the law has the form, this is what would happen (or this is what we would experience) *if* this is the case, where of course this is not and never can be the case. (The qualifications to such laws are often called "ceteris paribus" clauses—"other things being equal, then ...") This suggests that Newton's laws of motion are not derived from experience, or at least not in any straightforward fashion.

This being so, are they *a priori*? Were they just thought up by a Cartesian mind, a thinking thing, with no reference at all to what we get out of experience? This seems even more implausible, if it is taken to mean that a human mind could have thoughts even if it had never experienced an outside world at all. If we take seriously the idea that scientific laws are *a priori* in this sense, the implication is that a mind with *no* perceptual contact with anything outside of itself, excepting only God as the original implanter of innate ideas, will nevertheless have thoughts whose content is **applicable** to what there is. In the case of physical bodies such thoughts are limited to the bare possibilities of such bodies. That is, if there are physical bodies then they must be extended in space, capable of motion, rest and impact, and it must be possible to model these basic properties mathematically (more specifically, geometrically). On this view mathematics, including geometry, is *a priori*, as the *Meno* demonstrates.

This is not as absurd as it may seem. Socrates' doctrine of recollection is intended to show that if mathematical/geometrical truths are **innate**, it does not follow that I am aware of them. They need to be stimulated for me to recollect them. This is why Socrates casts himself in the role of an "intellectual midwife", particularly in the *Theaetetus* (3.1). In Descartes' case he denies that the mind is "transparent to itself"—that innate ideas are immediately there and obvious—and claims that it takes intellectual effort to recover or recollect them (5.2). The *Meditations* is, of course, intended as a self-help guide that the reader can use in order to arrive at these ideas.

If there are *a priori* truths, then, it does not follow immediately that they are present and correct in the mind in the absence of any external stimulus. It means only that once they are there in the mind—once they have been suitably triggered—it can be seen that they are independent of experience. When, for example, you learn geometrical truths about right-angled triangles, how is it that what you learn applies to *all* right-angled triangles? What is it that we grasp when we see that mathematical/geometrical principles apply in all relevantly similar cases? Do we generalise from a particular experience or set of experiences, or do we come to grasp truths of a different order, perhaps of something akin to a "**form** of triangle" that we have in mind?

These are difficult questions that do not have clear answers, which is why philosophers spend so much time on them. If we did have the answers they would cease to be philosophical questions and become the subject-matter of other disciplines; for the moment they remain puzzling. But this difficulty of getting clear about the exact function of experience is key to understanding that the differences between rationalism and empiricism are not as clear-cut as they may at first appear. Given this, here is a suggestive rule-of-thumb for deciding whether a principle is *a priori* or *a posteriori*: if a counter-example to a rule is observed, would you be prepared to change the rule, rather than find some other way to explain it away? If the answer is no, it is likely the rule is based on an ***a priori*** principle. If yes, it is most likely based on an ***a posteriori*** principle.

We can see this in the case of Newton's laws. If we observe a body moving in a straight line, and then see it deviate from that line, even if we cannot see why it deviates from that line we would look for a cause. It is hard to envisage any circumstances in which we would be prepared to consider changing the law. Similarly when accident investigators look into an aircraft crash we expect them to come up with a cause that fits with our present science, and if they fail to do so we assume that they have not found the cause, not that there is something wrong with our science (or that some supernatural event has occurred, that the laws of nature were momentarily suspended).

6.2 "Hypotheses non fingo"

It is extraordinary that from laws that give every impression of being *a priori*, Newton creates by mathematical techniques a physics that is **applicable** to our world of physical bodies. Descartes failed to do this, and in an obvious dig at his work Roger Cotes, the Plumian Professor of Astronomy and Experimental

Philosophy at Cambridge, a close associate of Newton and editor of the second edition of the *Principia*, says in his editor's preface that:

> Those who take the foundation of their speculations from hypotheses, even if they then proceed most rigorously according to mechanical laws, are merely putting together a romance, elegant perhaps and charming, but nevertheless a romance. (*Principia Mathematica*, p.386/p.43, Cohen/Whitman translation)

Hall says of Descartes' *Principles* that it "was a triumph of fantastic imagination which happens, unfortunately, never once to have hit on a correct explanation" (*From Galileo to Newton*, p.120), which is all too true. Towards the end of his preface Cotes says this:

> These [laws of nature] therefore we must not seek from uncertain conjectures, but learn them from observations and experiments. He who thinks to find the true principles of physics and the laws of natural things by the force alone of his own mind, and the internal light of his reason; must either suppose that the world exists by necessity, and by the same necessity follows the laws proposed; or if the order of nature was established by the will of God, that himself, a miserable reptile, can tell what was fittest to be done. All sound and true philosophy is founded on the appearances of things ...Fair and equal judges will therefore give sentence in favour of this most excellent method of philosophy, which is founded on experiments and observations. (*Principia Mathematica*, p.397–8/p.56–7)

But what are these observations and experiments? In discussing Descartes, two ways of reading the *Meditations* were proposed; the "defence of Cartesian physics" and the "response to scepticism" approaches (5.4). The suggestion made is that our approaches to texts—the phenomena, so to speak, in a humanities discipline like philosophy—are inevitably coloured by our preconceptions. This extends, literally, to what we see. It might be thought, initially, that it is absurd to claim that the earth is spherical; how is it that people at the antipodes do not suffer a rush of blood to the head before falling off? But observation of ships sailing out to sea disproves the idea that the earth is flat, because if it is then as a ship sails out to sea it should appear to get smaller and smaller until it is no longer visible. But this is not what happens; the ship appears smaller and smaller, certainly, but what happens is that the hull of the ship falls out of view first, until only the tops of the masts can be seen, because, of course, the earth is a sphere. (It could be a cylinder, but correlating observations of ships sailing away in different directions shows it is a sphere.) It is a capacity to see things differently that has often been the key to scientific discovery.

Something similar can be said about Newton and gravity. To see gravity at work in the fall of an apple or the orbit of the moon is to see things differently from Aristotle, whose views we will look at briefly next.

Excursus—Aristotle's Physics

In some ways this constitutes a digression, albeit a brief one, to look at the world-view that Descartes and Newton rejected. In this context there are three significant aspects of Aristotle's physics:

(i) the conceptual nature of his approach,
(ii) that action can only occur through contact,
(iii) the difference between the earthly and the heavenly (the subjects of physics and of *first philosophy*, or metaphysics, respectively).

Aristotle has a profound and thoroughly worked-out world view, so this is an extremely partial choice of topics, intended to fit with themes developed in the present work.

To begin with (i), in Aristotle's day there was none of the penumbra of apparatus and experimental techniques that we take for granted. There are no equations or mathematical formulae in Aristotle's *Physics* or in his *Metaphysics*. Both works consist in the main of conceptual arguments conducted in abstract fashion, and are, to put it mildly, boldly speculative. Here, for example, is Aristotle's argument for the primary nature of circular motion:

> As for the fact that circular movement is the primary kind of movement, this is obvious ...the higher degree of simplicity and completeness possessed by circular movement means

> that it has priority over [movement in a straight line]. In the first place, it is impossible to move over an infinite straight line, because there can be no such thing as a straight line which is infinite in this sense; also, even if there were such a thing, it would not be traversed by anything, because the impossible does not happen and it is impossible to traverse something which is infinite in extent. In the second place, movement on a finite straight line can either reverse direction or not; if it does, it is a composite of two movements, and if it does not, it is incomplete and must cease to exist. But where priority in nature, in definition, and in time are concerned, the complete is prior to the incomplete and that which does not cease to exist is prior to that which does. Besides, a movement that can be eternal is prior to one which cannot. Now, circular movement can be eternal, but no other kind of movement, and no other kind of change either, can be eternal, because they are bound to involve rest, and the presence of rest means that the movement or change ceased to exist. (*Physics*, p.265a13–26)

This displays a certain *a priori* philosophical style, as well as forming part of the distinction Aristotle draws between the earthly and the heavenly (iii, above). Movement on earth does not exhibit the eternal circular motion of the heavenly bodies, because the earth is the realm of the finite, of what comes to be and can be destroyed, whereas what is above the earth is eternal and neither comes to be nor ceases to be. First philosophy deals with these sorts of things, whereas physics or second philosophy deals with the perceptible things we interact with. (*Metaphysics*, p.1037a)

The claim that action can only occur through contact (ii) is a product of this conceptual approach. The idea of action at a distance seems fantastic; how can it be possible? Experience shows that things act on one another by impact, either directly in the case of hammers and axes and billiard balls and so on or indirectly, in the case of a varying series of pressure waves impinging on your eardrums leading you to act according to whatever warning or instruction you have heard.

Aristotle makes this point in *Physics*, Bk.VII, where he argues that:

> [T]he immediate agent of bodily change of place must be either in contact with or continuous with the moved object, as we always observe to be the case. So it necessarily follows that the moved objects and the movers are either continuous or in contact with one another. (*Physics*, p.242b59–61)

This is not an isolated view. It occurs elsewhere in the *Physics* (at p.202a3, 266b25) and in *On Coming-to-be and Passing-away* (at p.322b–323a). It leaves Aristotle with a particularly vexing problem, of what it is that keeps a projectile in motion (*Physics*, p.266b), because on his account it should stop moving as soon as a contact force ceases; as soon as a spear leaves a soldier's hand, for example. The contact doctrine forces him into a deeply implausible claim that motion is somehow transferred to and from the air (or water) that a projectile moves through, in a sequence of impulses that gradually die out. However implausible we may find this, the conceptual nature of the argument is significant. Given the assumption, that change requires contact, any perceptible change requires something in contact to bring it about, so something must be found to do the job. And Aristotle's intellectual prestige lent his arguments enormous weight.

A consequence of (iii) is that accounts of what happens on earth have no obvious relevance to what happens above the earth. The fall of an apple is no guide to the continual falling towards the earth exhibited by the Moon, a falling that keeps it in orbit (rather than flying off into space in a straight line). Aristotle's *Physics* operates with four elements—earth, air, fire, and water—each of which has a natural place that it tends towards. The fall of an apple is a natural motion whereby a heavy body tries to take up its proper place in the order of things, whereas the Moon as a heavenly body eternally sweeps out its circular motion. Newton's brilliance lay in seeing the same phenomena as everybody else—the motions of apples as well as those of the heavenly bodies—and realising that what makes the former fall, and keeps the latter in their orbits, might not be an intrinsic tendency on the part of heavy bodies to fall to earth or something in contact pressing them back from the outside but a force, albeit with no visible means of action, pulling them from the inside. Just as if you swing a weight round on a piece of rope you have to pull on the rope to stop it from flying off at a tangent.

Gravity

Newton's radical move was to propose a *centripetal* force that is equal and opposite to the *centrifugal* force that we're all familiar with. This centripetal force is, of course, the force of gravity. Newton realised that this can account for both sets of phenomena, the fall of apples and the orbits of the planets. Rather than attributing to apples a desire to rejoin bodies of their kind (i.e. bodies with mass), what makes them fall rather than float away is the attractive force of gravity. Similarly what prevents the Moon flying off into space is gravity as an attractive force between the earth and the Moon.

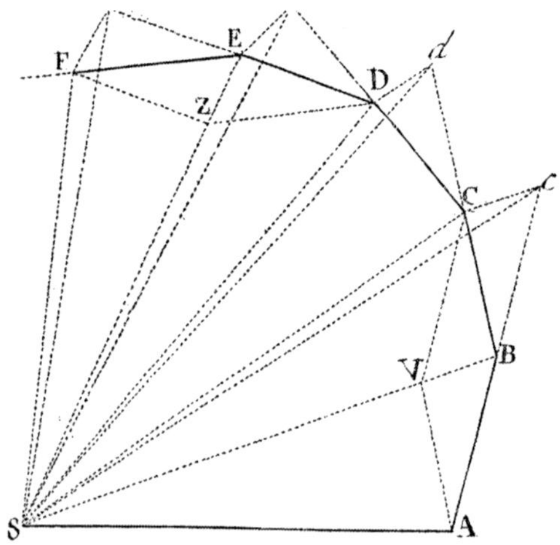

Centripetal forces

The diagram on the left is the first in Book 1 §2 of the *Principia*, entitled "To find centripetal forces". If S is the sun and A the earth, what stops the earth at B from flying off along the line Bc? Answer, a centripetal force, exerted along a celestial "piece of rope" with the sun at the other end. The "experiments and observations" that led to Newton proposing such a force took the form of astronomical data collected using telescopes. Some of this was available to Descartes, who saw what Newton saw, but was so taken by contact that he attributed free fall in the case of apples falling to earth or circular motion in the case of the planets to vortexes of "subtle matter" around the earth and the sun.

Descartes' approach relies on the **rationalist** way of accounting for our knowledge of what cannot be readily derived from experience. It is true that experience is required to bring out what is known innately, but the function of experience is only to prompt such bringing forth or recollecting. The problem for rationalism remains, though, of putting together our **innate** ideas (the ones we have in mind irrespective of anything external to us) and the world outside of us; why should it be the case that our innate ideas are appropriate for—are **applicable** to—this world? Why should not it be the case that our innate ideas are of chess but the world is a backgammon set? And if it is, would we ever realise? How could we ever come to modify and revise our theories, if we cannot step outside them to see that we are playing chess in a backgammon world?

To get over this, as we have seen (5.3), Descartes argues that a benevolent God has created us in such a way that our innate ideas are suitable for the world around us. This is less than convincing. To conclude this section, we will look at two passages from Newton's *Principia* that deal with observation and experiment and the nature of the rules and principles he puts forward. The first is from the "General Scholium", added to the second edition:

> Hitherto we have explained the phenomena of the heavens and of our sea, by the power of gravity, but we have not yet assigned the cause of this power. This is certain, that it must proceed from a cause that penetrates to the very centres of the sun and the planets, without suffering the least diminution of its force; that operates, not according to the quantity of the surfaces of the particles upon which it acts, (as mechanical causes [are wont] to do), but according to the quantity of the solid matter which they contain, and propagates its virtue on all sides, to immense distances, decreasing always in the duplicate proportion [square] of the distances, as far as the orb of Saturn ...
>
> But hitherto I have not been able to discover the cause of those properties of gravity from phenomena, and I frame no hypotheses [*hypotheses non fingo*]. For whatever is not deduced from the phenomena, is to be called an **hypothesis**; and hypotheses, whether metaphysical or physical, whether of occult qualities or mechanical, have no place in experimental philosophy. In this philosophy particular propositions are inferred from the phenomena, and afterwards rendered general by **induction**. Thus it was that the impenetrability, the mobility, and the impulsive force of bodies, and the laws of motion and of gravitation, were discovered. And to us it is enough, that gravity does really exist, and act according to the laws which we have explained, and abundantly serves to account for all the motions of the celestial bodies, and of our sea. (*Principia Mathematica*, p.943/p.92)

Newton's approach here exerted an enormous influence on **empiricist** philosophers and philosophers of science. The argument is that if the mathematics generates laws that serve as a predictive mechanism, that can be used to calculate the phenomena (for example the orbit of Saturn or the next return of Halley's Comet), then speculation about some deeper underlying causal mechanism can be discounted as metaphysical romancing. "Gravity" is a name for an observable, measurable phenomenon, whose effects can be calculated and predicted using mathematical models, and this is all that needs to be said about it. Science has no need for "feigned hypotheses" in such cases.

The second passage is from the "Rules for the Study of Natural Philosophy", as set out in the third edition of the *Principia*:

Rule 3 *The qualities of bodies, which admit neither intension nor remission [neither increase nor diminution] of degrees, and which are found to belong to all bodies within the reach of our experiments, are to be esteemed the universal qualities of all bodies whatsoever.*

For since the qualities of bodies are known to us by experiments, we are to hold for universal, all such as universally agree with experiments; and such as are not liable to diminution, can never be quite taken away. We are certainly not to relinquish the evidence of experiments for the sake of dreams and vain fictions of our own devising: nor are we to recede from the analogy of nature, which uses to be simple, and always consonant to itself. We no otherwise know the extension of bodies, than by our senses, nor do these reach it in all bodies; but because we perceive extension in all that are sensible, therefore we ascribe it universally to all others also.

That abundance of bodies are hard we learn by experience. And because the hardness of the whole arises from the hardness of the parts, we therefore justly infer the hardness of the undivided particles not only of the bodies we feel but of all others. That all bodies are impenetrable, we gather not from reason, but from sensation. The bodies which we handle we find impenetrable, and thence conclude impenetrability to be a universal property of all bodies whatsoever. That all bodies are moveable, and are endowed with certain powers (which we call the *vires inertiae* [forces of intertia]) of persevering in their motion or in their rest, we only infer from the like properties obeserved in the bodies which we have seen. The extension, hardness, impenetrability, mobility and *vis inertiae* of the whole, result from the extension, hardness, impenetrability, mobility, and *vires inertiae* of the parts: and thence we conclude the least particles of all bodies to be also all extended, and hard, and impenetrable, and moveable, and endowed with their proper *vires inertiae.* And this is the foundation of all [natural] philosophy ...

Lastly, if it universally appears, by experiments and astronomical observations, that all bodies about the earth, gravitate towards the earth; and that in proportion to the quantity of matter which they severally contain; that the moon likewise, according to the quantity of its matter gravitates towards the earth; that on the other hand our sea gravitates towards the moon; and all the planets mutually towards another; and the comets in like manner towards the sun; we must, in consequence of this rule, universally allow, that all bodies whatsoever are endowed with a principle of mutual gravitation. For the argument from the appearances concludes with more force for the universal gravitation of all bodies, than for their impenetrability; of which among those in the celestial regions, we have no experiments, nor any manner of observations. Not that I affirm gravity to be essential to bodies ...

Rule 4 *In experimental philosophy we are to look upon propositions collected by general* **induction** *from phenomena as accurately or very nearly true, notwithstanding any contrary hypotheses that may be imagined, till such time as other phenomena occur, by which they may either be made more accurate, or liable to exceptions.*

This rule we must follow that the argument of induction may not be evaded by hypotheses. (*Principia Mathematica,* p.795–6/p.87–9)

In Rule 3 Newton decisively rejects Aristotle's view of the heavens as divided into the sublunary sphere of change and decay, occupied by the earth and its inhabitants, and a celestial realm of unchanging eternal heavenly bodies in perpetual circular motion. As it is on earth, so to speak, so it is in the heavens. His emphasis on the role of experiment and observation is aimed squarely at the speculative philosophies of Aristotle and Descartes. He is careful to point out that he does not claim that "gravity is essential to bodies", which would amount to speculation about the **essence** of bodies; only that gravity is an observed, measurable phenomenon, because this is all he needs.

In Rule 4, he argues for the inductive method. The method of **induction** relies on accumulating observations, generalising the conclusions, and then arguing that it is probable that what has happened before will happen again. In the case of a counter-example, of course, the conclusions need to be modified unless some other way can be found of explaining the counter-example away. Until black swans were found in Australia it was thought (by induction) that the statement "all swans are white" is true. A sighting of a black swan falsified this statement. Closer examination might have shown that what appeared to be a black swan was a mutant duck, in which case the rule could be reinstated. Or it could have been argued that it was not really a swan at all, but such "ad hoc" strategies do not look to be good science.

6.3 Deduction and induction

The other technique for proof and argument is **deduction**. Deduction works by beginning with premises ($P_1, P_2, \ldots, P_n$) and then, by valid techniques, extracting conclusions (C). The underlying idea is that if the premises are true, whatever is validly derived from them must also be true; the rules of induction are *truth-preserving*. A deductive argument looks like this:

P_1 All swans are white.
P_2 Whatever is non-white is not a swan.
P_3 This swan is black.
C This "swan" is not a swan.

Deductive arguments protect their premises against putative counter-examples, as can be seen in this:

P_1	Every body perseveres in its state of being at rest or of moving uniformly straight forward, except in so far as it is compelled to change its state by forces impressed.
P_2	This body has deviated from its path.
P_3	No external force acting on the body has been detected.
C	There is an external force only we do not know what it is or how to detect it.

An inductive argument looks like this, with each premise (P_x) a record of an observation:

$P_1, P_2, \ldots, P_n$	This swan is white.
C_1	All swans are white.
P_{n+1}	This swan is black.
C_2	Swans are usually white but some are black.

The conclusion is modified in the light of the counter-example.

This is a contentious area in the philosophy of science generally, and what Newton meant by induction may not be exactly the same as our present notion. The interesting aspect here of Newton's work, which profoundly influenced later philosophers, is a deep tension between the *a priori* (what we know *a priori* and how we come to know it) and the *a posteriori* (what we know via the senses). In this sense Newton set the scene for the British empiricists (Locke, Berkeley and Hume), and for Kant.

Chapter 7

Locke's Empiricism; the Way of Ideas

Locke's justly famous *Essay Concerning Human Understanding* was published in 1689, two years after Newton's *Principia*, and is profoundly influenced by Newton's work. Locke's book is best known as a statement of the philosophical doctrine of **empiricism**, that what we know is derived in its entirety from experience. The mind of the new-born baby is akin to a blank slate that is impressed upon by what is outside of it (what is given in perception or sensation). These perceptions are taken up in the form of ideas, where by "idea" is meant broadly any content of the mind. Locke uses it as a technical term, so rather than try and offer a definition it is best to look at how he uses it.

If the mind merely took in what is perceived in the form of ideas it would be little more than a dumb recording machine. Consequently Locke also ascribes to the mind the power or capacity to reflect on its ideas, to order and arrange them. No new ideas are added in the course of ordering and arranging ideas, so it is in the ordering and arranging of ideas that we exhibit intelligence. There is a degree of psychological transparency to the mind. If we could observe a child from birth and make a note of all their experiences, we would have a record of all they know; the entirety of their stock of ideas. If we then set them an exam that involves comparing and contrasting their ideas, constructing arguments and drawing inferences from them, we could get a measure of intelligence, of their quickness and perspicacity in dealing with their ideas. What we would not find, according to

Locke, is any additional content, any additional **innate ideas** or principles, because he denies that there are any such things.

Of course this does not stop us from coming up with new theories. Locke could hardly propose a theory that makes Newton's achievements inexplicable and incomprehensible. But creative thinking on this approach is a matter of displaying unusual ingenuity in ordering and arranging ideas and in drawing more widely applicable consequences from them. Such theories have to be transparently explicable to others, accepting that not all will be sufficiently smart or have a sufficient background to understand a proposed theory. Nevertheless some degree of psychological transparency is a cornerstone of empiricism.

Locke begins his *Essay* with a first book (the first of four) in which he sets out to demolish the doctrine of innate ideas. I'm going to skip this because if you accept Locke's main argument for empiricism, you will not believe in innate ideas in any case. If you think there are matters that empiricism fails to account for (and we will touch on these later) then you are not going to buy Locke's main argument. This is a non-trivial aspect of studying different philosophical viewpoints. You rarely get a point-by-point rebuttal, more often than not philosophical views come as a package deal and buying one involves rejecting others as a matter of course. If someone likes soul and loathes hip-hop you're unlikely to convert them by enthusing over the virtuoso subtleties of your favourite artist.

7.1 Sensation and reflection

This is Locke's statement of his central claim:

> Every man being conscious to himself, that he thinks, and that which his mind is employed about whilst thinking, being the ideas, that are there, it is past doubt, that men have in their minds several ideas, such as those expressed by the words, *whiteness*, *hardness*, *sweetness*, *thinking*, *motion*, *man*, *elephant*, *army*, *drunkenness*, and others: it is in the first place then to be inquired, how he comes by them? I know it is a received doctrine, that men have native ideas, and original characters stamped upon their minds, in the very first being. This opinion I have at large examined already [in Book I of the *Essay*]; and, I suppose, what I have

said in the foregoing book, will be much more easily admitted, when I have shown, whence the understanding may get all the ideas it has, and by what ways and degrees they may come into the mind; for which I shall appeal to everyone's own observation and experience.

Let us then suppose the mind to be, as we say, *white paper*, void of all characters, without any ideas; how comes it to be furnished? Whence comes it by that vast store, which the busy and boundless fantasy of man has painted on it, with an almost endless variety? Whence has it all the materials of reason and knowledge? To this I answer, in one word, from *experience*: in that, all our knowledge is founded; and from that it ultimately derives itself. Our observation employed either about external, sensible objects; or about the internal operations of our minds, perceived and reflected on by ourselves, is that, which supplies our understandings with all the materials of thinking ...

First, our senses, conversant about particular sensible objects, do convey into the mind, several distinct perceptions of things, according to those various ways, wherein those objects do affect them: and as we come by those ideas, we have of *yellow*, *white*, *heat*, *cold*, *soft*, *hard*, *bitter*, *sweet*, and all those which we call *sensible qualities*, which when I say the senses convey into the mind, I mean, they from external objects convey into the mind what produces there those perceptions. This great source, of most of the ideas we have, depending wholly upon our senses, and derived by them to the understanding, I call SENSATION.

Secondly, the other fountain, from which experience furnishes the understanding with ideas, is the perception of the operations of our own minds within us, as it is employed about the ideas it has got; which operations, when the soul comes to reflect on, and consider, do furnish the understanding with another set of ideas, which could not be had from things without: and such are, *perception*, *thinking*, *doubting*, *believing*, *reasoning*, *knowing*, *willing*, and all the different actings of our own minds; which we being conscious of, and observing in ourselves, do from these receive into our understandings, as distinct ideas, as we do from bodies affecting

> our senses. This source of ideas, every man has wholly in himself; and though it be not sense, as having nothing to do with external objects; yet it is very like it, and might properly enough be called *internal sense.* But as I call the other SENSATION, so I call this REFLECTION, the ideas it affords being such only, as the mind gets by reflecting on its own operations within itself.
>
> By REFLECTION then, in the following part of this discourse, I would be understood to mean, that notice which the mind takes of its own operations, and the manner of them, by reason whereof, there come to be ideas of these operations in the understanding. These two, I say, viz. external, material things, as the objects of sensation; and the operations of our own minds within, as the objects of reflection, are, to me, the only originals, from whence all our ideas take their beginnings ...
>
> The understanding seems to me, not to have the least glimmering of any ideas, which it doth not receive from one of these two. External objects furnish the mind with the ideas of sensible qualities, which are all those different perceptions they produce in us: and the mind furnishes the understanding with ideas of its own operations.
>
> These, when we have taken a full survey of them, and their several modes, combinations, and relations, we shall find to contain all our whole stock of ideas; and that we have nothing in our minds, which did not come in, one of these two ways. Let anyone examine his own thoughts, and thoroughly search into his understanding, and then let him tell me, whether all the original ideas he has there, are any other than the objects of his senses; or of the operations of his mind, considered as objects of his reflection: and how great a mass of knowledge soever he imagines to be lodged there, he will, upon taking a strict view, see, that he has not any idea in his mind, but what one of these two have imprinted; though, perhaps, with infinite variety compounded and enlarged by the understanding. (*Essay*, II.1.1–5)

It would not be too much to say that everything else in the *Essay* follows from thinking through the consequences of this position.

7.2 Simple ideas

The first significant consequence of this is a tripartite division, into simple ideas and the two categories that we construct from these, which Locke calls "mixed modes" and "ideas of substances". Of simple ideas, Locke says this:

> Though the qualities that affect our senses, are, in the things themselves, so united and blended, that there is no separation, no distance between them; yet it is plain, the ideas they produce in the mind, enter by the senses simple and unmixed. For though the sight and touch often take in from the same object, at the same time, different ideas; as a man sees at once motion and colour; the hand feels softness and warm in the same piece of wax: yet the simple ideas thus united in the same subject, are as perfectly distinct, as those that come in by different senses. The coldness and hardness, which a man feels in a piece of ice, being as distinct ideas in the mind, as the smell and whiteness of a lily; or as the taste of sugar, and smell of a rose: and there is nothing can be plainer to a man, than the clear and distinct perception he has of those simple ideas; which being each in itself uncompounded, contains in it nothing but one uniform appearance, or conception in the mind, and is not distinguishable into different ideas.
>
> These simple ideas, the materials of all our knowledge, are suggested and furnished to the mind, only by those two ways above mentioned, viz. sensation and reflection. When the understanding is once stored with these simple ideas, it has the power to repeat, compare, and unite them even to an almost infinite variety, and so can make at pleasure new complex ideas. But it is not in the power of the most exalted wit, or enlarged understanding, by any quickness or variety of thought, to invent or frame one new simple idea in the mind, not taken in by the ways before mentioned: nor can any force of the understanding, destroy those that are there. (*Essay*, II.II.1–2)

This is a **foundationalist** approach to knowledge, because it assumes a primary or fundamental class of mental contents out of which all else is constructed (by reflection, in Locke's terminology). This, though, only gives us half the story. The other half of such approaches usually consists in the claim that what is primary is also *certain* or *indubitable*, a claim Locke does indeed go on to make:

> Our simple ideas are all real, all agree to the reality of things ...But though whiteness and coldness are no more in snow, than pain is; yet those ideas of whiteness, and coldness, pain, etc. being in us the effects of powers in things without us, ordained by our Maker, to produce in us such sensations; they are real ideas in us, whereby we distinguish the qualities, that are really in things themselves. For these several appearances, being designed to be the marks, whereby we are to know, and distinguish things, which we have to do with; our ideas do as well serve us to that purpose, and are as real distinguishing characters, whether they be only constant in fact, or else exact resemblances of something in the things themselves: the reality lying in that steady correspondence, they have with the distinct constitutions of real beings. But whether they answer to those constitutions, as to causes, or patterns, it matters not; it suffices, that they are constantly produced by them. And thus our simple ideas are all real and true, because they answer and agree to those powers of things, which produce them in our minds, that being all that is requisite to make them real, and not fictions at pleasure. For in simple ideas ...the mind is wholly confined to the operation of things upon it; and can make to itself no simple idea, more than what it has received. (*Essay*, II.XXX.2)

Note that Locke does not claim that our simple ideas exactly resemble what they are ideas of. All he is entitled to say is that our ideas are our ideas, that they are what we have in mind as a result of sensation. This leaves open the possibility that what we perceive, the ideas that we have in mind, are representations—literally, *re-presentations*—of the things themselves, where these representations are intermediaries between our minds and what there is. As if the mind is an inner theatre in which ideas appear on stage before the mind's eye. Locke's theory of perception is thus **realist** because what we perceive is derived from the things themselves, but is a form of **representative realism** rather than direct or **naïve realism**. We will come back to this in 10.3. What matters for the moment is that our simple ideas are just there, in the mind whether we choose them to be or not, and are caused by what there is. They are there whether we choose them or not in the sense that we see what we see, hear what we hear, and so on; beyond choosing to look this way rather than that, say, we have little or no control over our perceptions.

7.3 Complex ideas

Given this stock of simple ideas, the mind creates complex ideas by combining them. It is in such combination alone that our minds play an active or creative role. Locke identifies two categories of complex ideas, which he calls "mixed modes" and "ideas of substances". The first of these are entirely created by us and are not answerable to anything in nature. By contrast ideas of substance are associated with things around us, in ways that go to the heart of Locke's views. So I'll get the first out of the way quickly so we can move on to the more interesting (and more influential) views on ideas of substances. Locke makes the dependance of mixed modes on us clear in this passage:

> *Mixed modes*, are not only made by the mind, but made very arbitrarily, made without patterns, or reference to any real existence. Wherein they differ from those of substances, which carry with them the supposition of some real being, from which they are taken, and to which they are conformable. But in its complex ideas of mixed modes, the mind takes a liberty not to follow the existence of things exactly. It unites and retains certain collections, as so many distinct specific ideas, while others, that as often occur in nature, and are as plainly suggested by outward things, pass neglected without particular names or specifications. Nor does the mind, in these of mixed modes, as in the complex ideas of substances, examine them by the real existence of things; or verify them by patterns, containing such peculiar compositions in nature. To know whether his idea of adultery, or incest, be right, will a man seek it anywhere among things existing? Or is it true, because anyone has been witness to such an action? No: but it suffices here, that men

> have put together such a collection into one complex idea, that makes the archetype, and specific idea, whether ever any such action were committed in nature itself, or no ...
>
> Nobody can doubt, but that these ideas of mixed modes, are made by a voluntary collection of ideas put together in the mind, independent from any original patterns in nature, who will but reflect, that this sort of complex idea may be made, abstracted, and have names given them, and so a species be constituted, before any one individual of that species ever existed. Who can doubt, but the ideas of *sacrilege*, or *adultery*, might be framed in the mind of man, and have names given them; and so these species of mixed modes be constituted, before either of them was ever committed; and might be as well discoursed of, and reasoned about, and as certain truths discovered of them, whilst yet they had no being but in the understanding, as well as now, that they have but too frequently a real existence? (*Essay*, III.V.3, 5)

Broadly, mixed modes are notions used in moral philosophy, law, politics, and so on. Because we invent them, disputes arise when two people associate a different set of simple ideas with a mixed mode. Whether we agree or disagree in classifying someone's behaviour as an instance of jealousy, say, may be at issue if I associate a different set of simple ideas with the notion of jealousy to you. But for theory of knowledge it is the cases where we aim to think about and talk about the world around us that is of greater concern.

When Locke talks of ideas of substances what he has in mind is, broadly, our ideas of things that can be sorted into kinds. Cabbages and kings, people and horses and gold rings are all examples of substances. Inevitably Locke's view is prone to first-person bias, because he treats of *my* experience, whoever I happen to be. It follows that my idea of an elm may differ from yours, to the point that my idea of what an elm is may be so indistinct and inadequate that I point to a beech tree and call it an elm. If your idea is similarly inadequate and indistinct you may agree with me and we will both be wrong, or you may point out to me that I am wrong, that the leaf shape is that of a beech, not an elm. In which case I should revise my set of ideas appropriately and do better at tree identification in the future.

However much I learn about trees, though, can I ever be said to *know* what an elm or a beech truly is? I can certainly build up a stock of simple ideas derived from my experiences, and I may become expert in the life-cycles of trees, their cultivation and propagation, the diseases they are prone to, and so on. But what would it be to *know* what a tree, or anything else, for that matter, is *in itself*? Does an ability to classify and identify count as knowledge, or must there be something more? To get clearer about this question we need to consider two of Locke's most influential doctrines, his treatment of the primary/secondary qualities distinction that was touched on in 5.3, and his distinction between **real** and **nominal essences**.

7.4 Primary and secondary qualities

This is Locke's statement of the distinction:

> Whatsoever the mind perceives in itself, or is the immediate object of perception, thought, or understanding, that I call *idea*; and the power to produce any idea in our mind, I call *quality* of the subject wherein that power is. Thus a snowball having the power to produce in us the ideas of white, cold, and round, the powers to produce those ideas in us, as they are in the snowball, I call *qualities*; and as they are sensations, or perceptions, in our understandings, I call them *ideas*: which ideas, if I speak of sometimes, as in the things themselves, I would be understood to mean those qualities in the objects which produce them in us.
>
> Qualities thus considered in bodies are, first such as are utterly inseparable from the body, in what estate soever it be; such as in all the alterations and changes it suffers, all the force can be used upon it, it constantly keeps; and such as sense constantly finds in every particle of matter, which has bulk enough to be perceived, and the mind finds inseparable from every particle of matter, though less than to make it self singly be perceived by our senses. E.g., take a grain of wheat, divide it into two parts, each part has still solidity, extension, figure, and mobility; divide it again, and it retains still the same qualities; and so divide it on, till the parts become insensible, they must retain still each

> of them all those qualities. For division (which is all that a mill, or pestle, or any other body, does upon another, in reducing it to insensible parts) can never take away either solidity, extension, figure, or mobility from any body, but only makes two, or more distinct separate masses of matter, of that which was but one before, or which distinct masses, reckoned as so many distinct bodies, after division make a certain number. These I call *original* or *primary qualities* of body, which I think we may observe to produce simple ideas in us, viz. solidity, extension, figure, motion, or rest, and number.
>
> Secondly, such qualities, which in truth are nothing in the objects themselves, but powers to produce various sensations in us by their primary qualities, i.e. by the bulk, figure, texture, and motion of their insensible parts, as colours, sounds, tastes, etc. These I call the *secondary qualities.* To these might be added a third sort which are allowed to be barely powers though they are as much real qualities in the subject, as those which I to comply with the common way of speaking call *qualities*, but for distinction *secondary qualities.* For the power in fire to produce a new colour, or consistency in wax or clay by its primary qualities, is as much a quality in fire, as the power it has to produce in me a new idea or sensation of warmth or burning, which I felt not before, by the same primary qualities, viz. the bulk, texture, and motion of its insensible parts. (*Essay*, II.VIII.9–10)

The primary qualities of bodies, then, are solidity, extension, figure, motion or rest, and number. These are the properties of bodies that are susceptible to mathematical/geometrical treatment, that is, the properties of bodies that are relevant to mathematical physics. Locke's approach here is similar to that taken by Descartes, as we saw in 5.3.

But Locke differs in his view that *all* we know about bodies comes from experience. Whereas Descartes sets out an *a priori* mathematical physics based on *a priori* ideas concerning bodies as things that are extended in space, can move or be at rest, and can collide with one another (with consequent transfer of motion or rest), Locke can only derive such ideas from experience. This makes the notion of secondary qualities central to Locke's views because it is the secondary qualities of bodies that we perceive. It is these qualities that "are nothing in the objects themselves, but powers to produce various sensations in us by their primary qualities." It is from our perception of these secondary qualities of bodies that we infer their primary qualities. This is why Locke's theory of perception is a form of **representative realism**; what we perceive is real, but we perceive it in a mediated sense, through the ideas we form of things in terms of their secondary qualities. I see and smell a rose, and its thorn scratches my hand which produces red blood, but what there is in reality is just solidity, extension, figure, number, and so on; a veiled, colourless world that cannot be directly experienced (the **veil of perception**). I come to know about the external world because of the impact, literally, that it makes on my senses:

> The next thing to be considered, is how bodies produce ideas in us, and that is manifestly by *impulse*, the only way which we can conceive bodies operate in . . .let us suppose . . .that the different motions and figures, bulk, and number of such particles, affecting the several organs of our senses, produce in us those different sensations, which we have from the colours and smells of bodies, e.g. that a violet, by the impulse of such insensible particles of matter of peculiar figures, and bulks, and in different degrees and modifications of their motions, causes the ideas of the blue colour, and sweet scent of that flower to be produced in our minds . . .
>
> The particular bulk, number, figure, and motion of the parts of fire, or snow, are really in them, whether any one's senses perceive them or no: and therefore they may be called *real qualities*, because they really exist in those bodies. But light, heat, whiteness, or coldness, are no more really in them, than sickness or pain is in manna. ["Manna" is the miraculous food that sustained the Israelites in the wilderness.] Take away the sensation of them; let not the eyes see light, or colours, nor the ears hear sound; let the palate not taste, nor the nose smell, and all colours, tastes, odours, and sounds, as they are such particular ideas, vanish and cease, and are reduced to their causes, i.e. bulk, figure, and motion of parts. (*Essay*, II.VIII.11, 13, 17)

It is the primary qualities of bodies that are real, and these act on us by impulse and thereby cause our perceptions of bodies in terms of their secondary qualities. Causation is a matter of impulse, because this is the only way that the mechanical philosophy endorsed by Descartes and by Locke can account for relations between bodies—and perception is just that, a relation between bodies (and between my body and my mind, but for the sake of the argument we will skip this problem). Of course Locke knew about Newton's theory of gravitational attraction, but at the time Locke was composing the *Essay* the question of action at a distance was obscure. Newton's famous remark "hypotheses non fingo" appeared not in the first 1687 edition of *Principia Mathematica* but only in the second edition of 1713. It is striking that Locke says that impulse—contact—is "the only way we can *conceive* bodies operate in." Action at a distance occurs but remains, in a very real sense, inconceivable.

An immediate consequence of this is that bodies in terms of their primary qualities, how they are in themselves, are inaccessible to us. We cannot know how things really are, we can only know them in terms of causal interactions with our sensory organs, resulting in perceptions of (ideas of) things in terms of secondary qualities being deposited in the mind. The consequences of this will become clearer once Locke's second important distinction is in view, that between **real** and **nominal essence**.

7.5 Real and nominal essence

The **essence** of something is the properties that make it what it is. If it lacked these properties it would not be what it is, rather it would be something else. If, for example, the essence of water (what it is to be water) is to be made up of molecules consisting of two hydrogen atoms and one oxygen atom (H_2O), whatever is not so composed is not water. If it is of the essence of a swan to be white then a black swan is not a swan, it is another kind of bird.

The notions of **essence** and **substance** are closely related. "Substance" is a sortal concept because it is classificatory. A cabbage is a cabbage, a man is a man, a whale is a whale, and so on. As a first approximation a substance is such that the word for it can appear in a sentence either as subject or as predicate; "a man is a man" has the substance word "man" as subject and as predicate (a predicate is a word that stands for a quality or property). Of course "yellow is yellow" or "four is four" are perfectly good sentences, but they are not subject-predicate in form, they are identity statements. In philosophy grammar is often a useful guide.

Given that substances are sortal concepts, how do we decide whether a given object is an instance of a given concept? How many pages must a leaflet have to become a book? What makes a whale a mammal rather than a fish? Aristotle says that man is a rational animal, for example, but that might be a bit hard on someone with a degenerative disease that robs them of the capacity for thought. Given such borderline cases, on what should we base our classificatory decisions?

The temptation is to reach for the notion of essence and claim that we can define what it is to be an x, or a y, or a z. An x, we might say, is something that has the following properties, where in Locke's terminology these will of course have to be secondary qualities because these are the only ones we can perceive. This suggests that essence is down to us, yet it seems clear that we confront things as they are. This tension leads Locke to his real/nominal essence distinction:

> First, essence may be taken for the *very being* of any thing, whereby it is, what it is. And thus the real internal, but generally in substances, unknown constitution of things, whereon their discoverable qualities depend, may be called their ***essence***. This is the proper original signification of the word ...And in this sense it is still used when we speak of the essence of particular things, without giving them any name.
>
> Secondly, the learning and disputes of the schools, having been much busied about genus and species, the word "essence" has almost lost its primary signification; and instead of the real constitution of things, has been almost wholly applied to the artificial constitution of genus and species. It is true, there is ordinarily supposed a real constitution of the sorts of things; and it is past doubt, there must be some real constitution, on which any collection of simple ideas co-existing, must depend. But it being evident, that

> things are ranked under names into *sort* or *species*, only as they agree to certain abstract ideas, to which we have annexed those names, the essence of each genus, or sort, comes to be nothing but that abstract idea, which the general, or sortal ...name stands for. And this we shall find to be that, which the word "essence" imports, in its most familiar use. These two sorts of essences, I suppose, may not unfitly be termed, the one the *real*, the other the *nominal essence.*
>
> Between the nominal essence, and the name, there is so near a connection, that the name of any sort of thing cannot be attributed to any particular being, but what has this essence, whereby it answers that abstract idea, whereof that name is the sign.
>
> Concerning the real essences of corporeal substances ...there are if I mistake not, two opinions. The one is of those, who using the word "essence", for they know not what, suppose a certain number of those essences, according to which, all natural things are made, and wherein they do exactly every one of them partake, and so become of this or that species. The other, and more rational opinion, is of those, who look on all natural things to have a real, but unknown constitution of their insensible parts, from which flow those sensible qualities, which serve us to distinguish them one from another, according as we have occasion to rank them into sort, under common denominations. (*Essay*, III.III.15–17)

Locke offers a tripartite account of **essence**:

(1) there are **nominal essences**, which are little more than names we assign to collections of simple ideas,
(2) there are **real essences** akin to Platonic forms that we cannot know, and
(3) there are **real essences** as consequences of the primary qualities of bodies (which we can infer something about from their secondary qualities, which we can perceive).

When we classify this animal as an elephant or this lump of metal as gold, are we classifying it in terms of nominal essence, or in terms of one or other of Locke's real essences? To backtrack for a moment, what matters to us is *classification*, where this is to be in accord with what there is. Assuming that there is an external world and it contains instances of substances, we want to be able to distinguish real gold from some inferior yellow metal substitute. What would count as a criterion to do this? First of all, we assume that there is some way of drawing such a distinction, it is just a matter of finding it. If there is no difference then there is no difference, if this piece of yellow metal is indistinguishable by any means available then it is gold and that is all there is to say about it. Of course some future chemical theory may show that it is not gold and that we are wrong, but all this shows is that the claims made by empiricists are **defeasible**, that they may be shown later to be wrong. Locke leaves this possibility open (it is bound up with the two approaches he suggests to real essences in the passage above; in the first case, because it is the approach of those who do not really know what they're talking about, in the second case, because it is the approach taken by those who know that they do not know all there is to know about what it is they're talking about).

The difficulty with nominal essence is that it fails to secure our intuitions about what there is. Nominal essences just seem too arbitrary and too prone to abuse from positions of power. What can stop a dictator from keeping all the "real" gold in a country back for himself and forcing his subjects to pay the gold price for jewellery made from a yellow metal substitute? What is there to stop him from legislating that his "fool's gold" is actually the real thing, if its everyday properties are indistinguishable from gold and he has a monopoly on all the laboratory tests that can tell the difference? In such circumstances nominal essences are rather too close to the "mixed modes" that we invent for ourselves.

This leaves his two categories of real essences, that essences are akin to Platonic forms, or that they arise in our minds via the secondary qualities that the primary qualities of relevantly similar objects cause in our minds. But in either case, real essences are not the sorts of things we can perceive, because we can only perceive things in terms of their secondary qualities, and we can only form complex ideas of things by collecting together the instances of the secondary qualities that we perceive in regular and repeated fashion. Here Locke finds himself with, philosophically speaking, a not uncommon problem—that of wanting to say something about certain kinds of things only to find that his own theory prevents him from doing so. Since his theory says that we cannot perceive real essences, in either sense, we can hardly get to know them. This Locke makes clear:

> If you demand, what those real essences are, it is plain men are ignorant, and know them not. From whence it follows, that the ideas they have in their minds, being referred to real essences as to archetypes which are unknown, must be so far from being adequate, that they cannot be supposed to be any representation of them at all. The complex ideas we have of substances, are ...certain collections of simple ideas, that have been observed or supposed constantly to exist together. But such a complex idea cannot be the real essence of any substance; for then the properties we discover in that body, would depend on that complex idea, and be deducible from it, and their necessary connection with it be known; as all properties of a triangle depend on, and as far as they are discoverable, are deducible from the complex idea of the three lines, including a space. But it is plain, that in our complex ideas of substances, are not contained such ideas, on which all the other qualities, that are to be found in them, do depend.
>
> The common idea men have of iron, is a body of a certain colour, weight, and hardness; and a property that they look on as belonging to it, is malleableness. But yet this property has no necessary connection with that complex idea, or any part of it: and there is no more reason to think, that malleableness depends on that colour, weight, and hardness, than that of that colour, or that weight depends on its malleableness. And yet, though we know nothing of these real essences, there is nothing more ordinary, than that men should attribute these sorts of things to such essences. The particular parcel of matter which makes the ring I have on my finger, is forwardly, by most men, supposed to have a real essence, whereby it is gold; and from whence those qualities flow, which I find in it, viz its peculiar colour, weight, hardness, fusibility, fixedness, and change of colour upon a slight touch of Mercury, etc. This essence, from which all these properties flow, when I inquire into it, and search after it, I plainly perceive I cannot discover: the furthest I can go, is only to presume, that it being nothing but body, its real essence, or internal constitution, on which these qualities

> depend, can be nothing but the figure, size, and the connection of its solid parts; of neither of which, I having any distinct perception at all, I can have no idea of its essence, which is the cause that it has that particular shining yellowness; a greater weight than any thing I know of the same bulk; and a fitness to have its colour changed by the touch of Mercury. (*Essay*, II.XXXI.6)

Locke rejects the claim that **real essences** are Platonic forms, but in truth he has no sound reason for this because we cannot know real essences; if they cannot be known then there is no way to say what they are not. Whatever real essences are, the only claim he can reasonably make about them is that the primary qualities of things cause our ideas of these things in terms of their secondary qualities. In the act of classification we bring together relevantly similar bundles of co-located simple ideas, thereby getting as close as we can to real essences.

This is a critical point in the development of **empiricism**. Locke associates Platonic forms with **rationalism**, with the idea that if we know the **forms** of what there is, we can deduce from these all that there is to know about the behaviour of things (in the form of a **deductive** argument, as 6.1). All the properties of bodies would then be **necessary** properties, because they are essential to whatever it is. We would know all that there is to know about cause and effect, and by the same token we would know all the laws of nature; all true general propositions of the form "all As are Bs" or "all cases of A are followed by (cause) cases of B." Consequently rationalism tends to suffer from what we can call the **omniscience problem**, because it tends to ascribe to us knowledge that we do not and cannot obviously have. This is why Locke denies that our complex ideas of a substance can be of its **real essence**, because if it was "then the properties we discover in that body, would depend on that complex idea, and be deducible from it, and their necessary connection with it be known."

Locke's approach is rather by experiment and **induction**, but at bottom he is profoundly sceptical about the extent of human knowledge, in his own time and into the future:

> In some of our ideas there are certain relations, habitudes, and connections, so visibly included in the nature of the

ideas themselves, that we cannot conceive them separable from them, by any power whatsoever. And in these only, we are capable of certain and universal knowledge. Thus the idea of a right-angled triangle necessarily carries with it an equality of its angles to two right ones. Nor can we conceive this relation, this connection of these two ideas, to be possibly mutable, or to depend on any arbitrary power, which of choice made it thus, or could make it otherwise. But the coherence and continuity of the parts of matter; the production of sensation in us of colours and sounds, etc. by impulse and motion; nay, the original rules and communication of motion being such, wherein we can discover no natural connection with any ideas we have, we cannot but ascribe them to the arbitrary will and good pleasure of the wise architect ...

The things that, as far as our observation reaches, we constantly find to proceed regularly, we may conclude, do act by a law set them; but yet by a law, that we know not: whereby, though causes work steadily, and effects constantly flow from them, yet their connections and dependencies being not discoverable in our ideas, we can have but an experimental knowledge of them. From all which it is easy to perceive, what a darkness we are involved in, how little it is of being, and the things that are, that we are capable to know. And therefore we shall do no injury to our knowledge when we modestly think with our selves, that we are so far from being able to comprehend the whole nature of the universe, and all the things contained in it, that we are not capable of a philosophical knowledge of the bodies that are about us, and make a part of us: concerning their secondary qualities, powers, and operations, we can have no universal certainty. Several effects come every day within the notice of our senses, of which we have so far sensitive knowledge: but the causes, manner, and the certainty of their production, for the two foregoing reasons, we must be content to be ignorant of. In these we can go no further than particular experience informs us of matter of fact, and by analogy to guess what effects the like bodies are, upon other trials, like to produce. But as to a perfect science of natural bodies, (not to mention spiritual beings,) we are, I think, so far

> from being capable of any such thing, that I conclude it lost labour to seek after it. (*Essay*, IV.III.29)

The primary/secondary qualities distinction entails that we can only ever perceive what there is in terms of secondary qualities, in terms of the ideas that are caused in us by the actions of bodies on us. How things are in themselves is not something that we can ever know. But does it make sense to ask how things are in themselves? Why should not we say that they are as we perceive them, and that is all there is to say about it? Does it make sense to think that there is more? Or is Locke's pessimism about scientific progress, into the discovery of the inner workings of what there is, just mistaken?

In truth, **empiricism** is prone to the mirror image of the **omniscience problem** that tends to infect **rationalism**. Whereas rationalists tend to claim more knowledge than they have warrant for, empiricists tend to find that there are things that we know (or at least think we know) that cannot readily be founded on experience (sensation and reflection) alone. But this is all perhaps rather abstract, so an illustration drawn from Descartes' and Newton's views on the movements of the heavenly bodies may be useful.

Chapter 8

Case Study 1: Descartes and Newton on the Movements of the Heavenly Bodies

It is all too easy in philosophy to get bogged down in abstract disputes that do not seem to have much connection with reality. This applies as much to theory of knowledge as to any other area of the subject. Rather than just talk about knowledge and the differences between rationalism and empiricism in the abstract, in this chapter and the next we will look at concrete examples of how these two approaches operate.

The subject of the movements of the heavenly bodies was a popular topic in the seventeenth century. In his *Dialogue Concerning the Two Chief World Systems*, published in 1632, Galileo suggested that the earth moves. What today seems a thoroughly innocuous statement was, then, sufficient to incur the wrath of the Pope and the censure of the Catholic Church. The first system is the Ptolemaic system, after the ancient Greek astronomer Ptolemy, which assumes that the earth is a fixed point at the centre of the universe. The second system, due to Copernicus (1478–1563), holds that the sun is at the centre of our solar system and the earth orbits around it. Surprising though it may seem, in Galileo's time there was no clear way of distinguishing between them—there is no *observable* difference—but Catholic orthodoxy held as an article of faith, not open to scientific investigation, that the earth is at the centre of the universe. It was permissible to put forward the idea that the sun is at the centre with the earth and the other planets revolving around it as a **hypothesis**—a hypothesis that is false, of course—but no more than that. Quite why one would want to work with a false hypothesis is another matter, as it rather undermines the point of an argument, whether inductive or deductive, but perhaps religion and science do not always go together terribly well.

Galileo had been warned before 1632 not to propound the idea that the earth moves, but the Pope and the Inquisition felt he had gone too far in his *Dialogue* towards promoting the idea that the earth moves as a claim rather than as a (false) hypothesis. Galileo was condemned and put under house arrest. His book was put on the index of banned works, and copies of it were withdrawn and burned.

It was against this background that Descartes suppressed a book called *The World* which he was working on in 1630/33, because in it he assumes that the earth revolves around the sun. Descartes reworked most of the book into his *Principles of Philosophy* (1644), which Newton studied closely.

The aim in this section is to illustrate Descartes' method and the remarkable conclusions it led him to. The method is to propose first principles which are "clear and distinct", in Descartes' terms. They have to meet a certain standard that makes them indubitable. As we have seen (at 5.2) Descartes does not hold that the mind is transparent to itself, so we have to discover or recollect these first principles. Crucially these principles are ***a priori***, that is, they are independent of experience, where independence is explained by the claim that they are **innate**.

With these first principles in place, Descartes deduces consequences from them (as 6.3). Given that the first principles do indeed meet the criteria established, and that the consequences are **validly** deduced (according to rules that are truth-preserving), the consequences have to be true, however outlandish they may be. This is the **rationalist** method at work. In the next section we will look at the extraordinary theory it led Descartes to propound concerning the movements of the heavenly bodies.

8.1 Descartes' vortex theory

The rationalist begins by establishing first principles. These are to be set up by pure thought alone, independently of experience (that is, they are *a priori*). As for their origin, they can be thought of as innate, or as recovered by recollection, perhaps by some form of intellectual exercise along the lines of the *Meditations*. If we follow the method of the *Meditations*, whatever we come up with in pure thought has to meet certain criteria if it is to be accepted as a first principle; it can hardly be that whatever comes to mind is suitable.

Descartes proposes as criteria that our ideas are "clear" and "distinct":

> It is certain, however, that we will never mistake the false for the true provided we give our assent only to what we clearly and distinctly perceive . . . A perception which can serve as the basis for a certain and indubitable judgment needs to be not merely clear but also distinct. I call a perception "clear" when it is present and accessible to the attentive mind—just as we say that we see something when it is present to the eye's gaze and stimulates it with a sufficient degree of strength and accessibility. I call a perception "distinct" if, as well as being clear, it is so sharply separated from all other perceptions that it contains within itself only what is clear. (*Principles of Philosophy*, §§43, 45)

With this in place, he applies it to the ideas of physical bodies that come to him in the course of the *Meditations*. This leads him to a number of first principles, as follows:

(1) The essence of physical bodies is that they occupy (are extended) in space.
(2) There cannot be an instance of a vacuum in nature; nothing is literally nothing.
(3) Bodies can only act on one another by impulse (impact or collision) (as 6.2).
(4) God created the universe with a certain amount of motion, and this motion is conserved.

From (1) and (2), Descartes infers the following:

(I_1) Whatever seems to be empty space is in fact a body of some sort, bounded by other bodies.
(I_2) Space is full (usually stated as "space is a plenum" from the Latin *plenus*, meaning *full*).

Given these, Descartes infers that:

(I_3) When a body moves into a new location it displaces the body in that location.
(I_4) All motion is circular, because a body has to move into the location vacated—all motion has to be in a completed loop.

Descartes thinks that our ideas of these principles meet the criteria of clearness and distinctness, so any account of the phenomena that we experience has to be explicable in terms of them. This is the order the **rationalist** works in; begin with first principles that are warranted by meeting suitable criteria, independently of experience, and then account for the phenomena (what is experienced) in terms derived from first principles. The phenomena to be accounted for here are the motions of the heavenly bodies, and for the sake of the argument it is assumed these follow broadly circular orbits irrespective of where the centre of the universe is taken to be (the earth or the sun, Ptolemy or Copernicus respectively).

If you take a bucket of water, tie a rope to its handle and then swing it round in a circle, you very obviously have to exert some force on the rope to keep the bucket in a circle. If you let go of the rope the bucket flies off in a straight line, at a tangent to the circle. In this instance it is your effort in holding on to the rope that keeps the bucket moving in a circle. But in the case of our solar system (assuming that the sun is at the centre of our planetary system), what keeps the earth and all the other planets in their orbits, and the Moon and the satellites of Jupiter and Saturn in their orbits; what stops them from flying off into space? Where, so to speak, are all the pieces of rope?

Given the principles set out above, Descartes has to give an account that relies on impulse, on direct action through contact between bodies. Literally there has to be *something* that acts on each of the planets to keep them in their orbits, but what? Given that there is nothing visible, whatever it is has to be invisible. So Descartes proposes a fine, very fluid (much more fluid than water

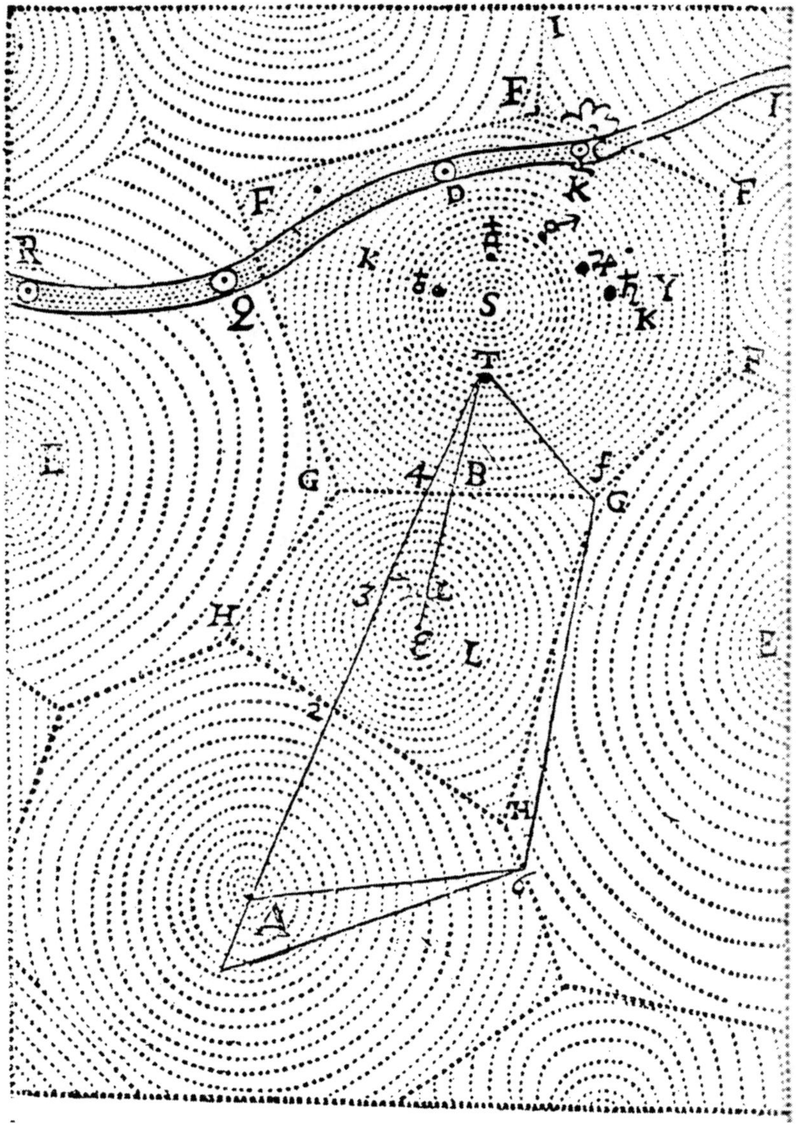

The universe, as depicted in Descartes' *Treatise on Light* (1664). S is our sun, T the earth, with the irregular hexagon around them representing the **vortex** system of our universe. The other stars each have their own system of vortexes. The "snake" from right to left is the passage of a comet.

at room temperature) substance that fills the heavens. This fluid rotates in vortexes around the sun, carrying the planets along with it and constantly pressing them back into their orbits. It is far more fluid than the planets, which appear solid to us, and is fluid enough to allow an even finer substance—light—to pass through it. (It is the pressure exerted by this even finer substance on our eyeballs that we experience as seeing.)

Each star will have its own set of planets, so the universe is made up of a series of individual solar systems. Descartes hypothesises that comets must be the most massive and solid bodies there are, because what we would now think of as their momentum means that the force exerted on them by the vortexes is insufficient to hold them within a single solar system. (In the absence of a theory of gravity, Descartes did not have a clear distinction between mass (the quantity of "stuff" a body is composed of) and its weight (mass × the force of gravity acting in its locality). Since momentum = mass × velocity, Descartes did not have the concept of momentum as we now have it, but he clearly had a somewhat analogous concept.) Quite how comets achieve circular motion is not clear, and they form an obvious counter-example to Descartes' theory.

However matters might lie with comets, though, in accord with (I_4), above, Descartes' **vortex theory** was a serious attempt to account for the phenomena—the motions of the heavenly bodies—by inference from a set of *a priori* first principles. But given that these first principles are *a priori*, that is, are independent of experience, and given that the usual pattern of a deductive argument is to protect what is deduced from any form of refutation, how can an *a priori* theory like Descartes' vortex theory be overturned and be shown to be wrong? From rationalism, then, we turn to the inductive, experimental approach taken by Newton.

8.2 Newton and action at a distance

If we set aside Descartes' first principles, we are left with a phenomenon to be explained; that the planets do indeed move in roughly circular orbits. Setting aside these first principles does not mean that we have no basis whatever to theorise upon. We have all the evidence we can derive from observation, with the naked eye and with telescopes. Even in the seventeenth century a wealth of historical astronomical data was available. The bucket on a string model leaves open the idea that there is something that pushes the planets back into their orbits from the outside, or that there is something that pulls them from the inside. We can still put forward first principles as working hypotheses. The difference in going about matters this way is that we begin with the phenomena, as a consequence of which whatever principles we put forward are **defeasible**, that is, we are prepared to discard or modify them as needs be. Theorising becomes a dynamic interplay between principles and experience, and this is one of the strong points of **empiricism**.

Given the phenomena, Newton begins with the observation that there must be some *way* in which the planets are kept in their orbits, leaving it open whether there is some *thing* that does the job. This allows Newton to suspend the idea that the only interactions there can be between bodies consist in impulse, enabling him to consider a model in which the phenomena are as they are whether or not we can provide an explanation of them. The phenomena can be investigated mathematically irrespective of how they come about. In the "Editor's Preface to the Second Edition" of the *Principia*, approved by Newton, Cotes says this:

> Those who have treated of natural philosophy, may be nearly reduced to three classes. Of these some have attributed to the several species of things, specific and occult qualities; on which, in a manner unknown, they make the operations of the several bodies to depend. The sum of the doctrine of the Schools derived from Aristotle and the Peripatetics is herein contained. They affirm that the several effects of bodies arise from the particular nature [**essences**] of those bodies. But whence it is that bodies derive those natures they don't tell us; and therefore they tell us nothing. And being entirely employed in giving names to things [i.e. Lockean nominal essences, as (1) at 7.5], and not in searching into things themselves, we may say that they have invented a philosophical way of speaking, but not that they have made known to us true philosophy.
>
> Others therefore by laying aside that useless heap of words, thought to employ their pains to better purpose. These supposed all matter homogeneous, and that the variety of forms

> which is seen in bodies arises from some very plain and simple affections of the component particles. And by going on from simple things to those which are more compounded they certainly proceed right; if they attribute no other properties to those primary affections of the particles than nature has done. But when they take a liberty of imagining at pleasure unknown figures and magnitudes, and uncertain situations and motions of the parts; and moreover of supposing occult fluids, freely pervading the pores of bodies, endowed with an all-performing subtlety, and agitated with occult motions [i.e. Descartes' "fine substance", as 8.1]; they now run out into dreams and chimeras, and neglect the true constitution of things; which certainly is not to be expected from fallacious conjectures, when we can scarce reach it by the most certain observations. Those who fetch from hypotheses the foundation on which they build their speculations, may form indeed an ingenious romance, but a romance it will still be.
>
> There is left then the third class, which professes experimental philosophy. These indeed derive the causes of all things from the most simple principles possible; but then they assume nothing as a principle, that is not proved by phenomena. They frame no hypotheses, nor receive them into philosophy otherwise than as questions whose truth may be disputed. (*Principia Mathematica*, p.385–6/p.43)

Descartes belongs to the second class, Newton to the third. In his book Newton adopts a strategy of demonstrating from what was known about the motions of bodies and fluids that **vortex theory** cannot account for the observed facts; for example, what happens when one system of vortexes impinges on a second, as in Descartes' diagram at p.109, given that space is a plenum? Indeed Book II of *Principia Mathematica* is devoted to fluid mechanics, to show that results derived from work on fluid flow discredit vortex theory as an account of the movements of the heavenly bodies. Given that an individual Cartesian solar system consists of vortexes as concentric shells, it is spherical in shape. But packing these into a universe leaves spaces, like packing oranges into a carton. Since space is a plenum (I_2), each individual solar system must be distorted to fill the gaps, as in the diagram. But this will lead to turbulent interference between adjacent solar systems, which will upset the orbits of the planets. (If they are partitioned from one another, the partitions would have to be transparent and permeable to comets but not to vortexes. The more *ad hoc* rules introduced, the less plausible the theory.) Evidence from observations and experiments on earth show that Descartes' theory fails to fit the facts. Newton goes on to argue that if gravity is accepted to be an attractive force between massive bodies, a seamless mathematical theory that is consistent with the observational data follows.

This is not, though, an *explanation* of what gravity is and how it comes about, and this led Leibniz, among others, to argue that Newton's gravity is a mysterious occult force. Consequently it is a backward step from vortex theory, because at least vortex theory has a readily understandable story to tell about the interactions of bodies (c.f. *Isaac Newton, Philosophical Writings*, p.111–17). Vortex theory is, of course, a product of the commonsense view, dating back to Aristotle (6.2) that causal relations can only be explained on the basis of impact. Positing gravity as a theoretical entity is justified by the predictive power of the mathematical theory it is embedded in, but this leaves our commonsense intuitions far behind. In fact Newton rejected the notion of action at a distance without an intervening medium as absurd (*Isaac Newton, Philosophical Writings*, p.102–3), but this leaves a gap to be filled.

In terms of knowledge the arguments revolve around questions of limits to what we can know, and consequently what we can explain, and the attitude we should take towards such limits. **Rationalists** tend to be motivated by the idea that everything is in principle knowable and that ultimate explanations from first principles are possible. By contrast **empiricists** tend to a more pragmatic acceptance that such explanations may not be possible, because science just does not work this way. Whereas rationalism suggests a static conception of a body of knowledge to be set out in an ordered way—a comprehensive theory of everything—empiricism suggests a dynamic process of formulating, testing and revising **defeasible** theories, a process that may never come to an end.

If you think of a universe in which the planets in our solar system whip round the sun at astonishing speed, and our solar system itself as in constant motion, what should we think of how this comes about? Is gravity a mysterious unseen force, a label to stick over something we cannot begin to explain—even though we can measure it and treat it mathematically? Should we stop at gravity or look for a deeper explanation, and if so, what would count as an explanation? Or is there a point at which explanations stop and we are left with brute facts?

There is another context within which it is interesting to think about these questions, that of religious, social and political values. Empiricism has become the dominant way of thinking in western secular societies. Its approach is essentially pragmatic and technical, and it constrains speculative thinking within certain limits, bounded by what is demonstrable, repeatable, and subject to mathematical methods. By contrast the rationalist temper is speculative and comparatively unconstrained, is more obviously rooted in religious ways of thinking, and is much less inclined to accept limits on what we can and cannot know. The theory of knowledge is very far from being a dry, technical subject in which all the questions can be solved by argument and demonstration.

Chapter 9

Case Study 2: Locke and Newton on the Very Small

9.1 Locke on substance

As we saw in 7.5, in looking at Locke's views on real and nominal essences, Locke claims that we can never get at the ultimate constituents of what there is, at substances themselves (gold, tigers, lemons or beeches, for example). But it is an open question whether Locke makes this claim because:

- we can only perceive things in terms of their secondary qualities, whereas the things themselves are the primary qualities, so they are unknowable by us, or;
- science will never be able to get at the ultimate constituents of what there is.

Note that these are consistent with one another; the interesting question is whether Locke's motivation is *philosophical* (in principle we can never know substances in terms of their primary qualities) or whether it is *scientific* (the science of Locke's day could not get at substances). It is not entirely clear which of these Locke holds. Following on from the preceding case study we can now go a bit further into this, based again on the differing approaches taken by rationalists and empiricists.

It is perhaps unfortunate to talk about rationalism and empiricism as if it is a dispute between philosophers rather than one between scientists. As a consequence of institutional boundaries, philosophers and scientists now tend to work at opposite ends of universities and have little contact with one another; philosophy is usually located with the humanities, alongside literature and languages, whereas science is science, with its own domain both physical and intellectual. Up until perhaps the second quarter of the nineteenth century, though, there was no such clear-cut demarcation, and all of Descartes, Newton, Locke, Berkeley, and the philosophers we will be looking at next—Hume and Kant—considered themselves to be students of natural philosophy. In thinking about what there is and how we get to know it all the people mentioned above were engaged in the same subject, each assuming that his way of doing the subject is the right way.

To revert to Locke, this is a brief extract from a passage quoted in 7.5:

> The complex ideas we have of substances, are ...certain collections of simple ideas, that have been observed or supposed constantly to exist together. But such a complex idea cannot be the real essence of any substance; for then the properties we discover in that body, would depend on that complex idea, and be deducible from it, and their necessary connection with it be known. (*Essay*, II.XXXI.6)

Following from 7.5, it can be seen that Locke's rejection of real **essences**—essences as Platonic **forms**, of type (2), as 7.5—is a rejection of rationalism. Knowledge of a substance (gold, for example), for a rationalist, would entail being able to **deduce** from a knowledge of its real essence all its secondary qualities, because there is a **necessary** connection between real essence and secondary qualities (the qualities that we perceive; that gold is yellow, heavy, malleable, and so on). If this is what is required for knowledge then, Locke thinks, we can never know substances in terms of real essence. We can only ever know something as we perceive it, that is, we can only ever know something in terms of its secondary qualities. But this leaves Locke in a bind, because his take on rationalism forces him to deny that we can ever get to know substances in any very deep sense by scientific investigation (c.f. Locke, *Essay*, IV.III.29 quoted at 7.5). In true pragmatic empiricist fashion, Locke hedges his bets:

> But whilst we are destitute of senses acute enough, to discover the minute particles of bodies, and to give us ideas of their mechanical affections, we must be content to be ignorant of their properties and ways of operation; nor can we be assured about them any further, than some few trials we make, are able to reach. But whether they will succeed again another time, we cannot be certain. (*Essay*, IV.III.25)

To be fair to Locke, in his day experiments often yielded inconsistent and unrepeatable results because of the difficulties of fixing conditions. In experimenting generally the idea is to fix certain conditions as a background and then, against this constant background, vary whatever it is you want to test. The difficulty in Locke's day of identifying pure chemical compounds meant that experiments were like solving multiple equations simultaneously, with too many variables. However Locke's approach suggests that he thinks there is more to our lack of knowledge of substances than merely practical difficulties in fixing background conditions and reducing the number of variable factors to manageable levels. But we do not yet have a compelling argument to show that scientific investigation of things is misguided. "Substance" is a philosophical notion that may prove irrelevant to science.

9.2 Newton on the cohesion of bodies

If gravity takes care of the forces between the very big, what of the very small? In the Preface to the First Edition of the *Principia*, published two years before Locke's *Essay*, Newton says this:

> For I am induced by many reasons to suspect that [the phenomena of nature] may all depend upon certain forces by which the particles of bodies, by some causes hitherto unknown, are either mutually impelled towards each other and cohere in regular figures, or are repelled and recede from each other; which forces being unknown, philosophers have hitherto attempted the search of nature in vain. (*Principia Mathematica*, p.382–3/41)

Newton returned to this question in query 31 of his *Opticks*, published in the 1730 edition, three years after his death:

> For we must learn from the phenomena of nature what bodies attract one another, and what are the laws and properties of the attraction, before we inquire the cause by which the attraction is performed. The attractions of gravity, magnetism, and electricity, reach to very sensible distances, and so have been observed by vulgar eyes, and there may be others which reach to so small distances as hitherto escape observation; and perhaps electrical attraction may reach to such small distances, even without being excited by friction ...
>
> The parts of all homogeneal hard bodies which fully touch one another, stick together very strongly. And for explaining how this may be, some have invented hooked atoms, which is begging the question; and others tell us that bodies are glued together by rest, that is, by an occult quality, or rather by nothing; and others, that they stick together by conspiring motions, that is, by relative rest amongst themselves. I had rather infer from their cohesion, that their particles attract one another by some force, which in immediate contact is exceeding strong, at small distances performs the chemical operations above mentioned, and reaches not far from the particles with any sensible effect ...
>
> And how such very hard particles which are only laid together and attach only in a few points, can stick together, and that so firmly as they do, without the assistance of something which causes them to be attracted or pressed towards one another, is very difficult to conceive. (*Opticks*, p.376, 388–9, 390/132, 133; in Newton's day "vulgar" meant "everyday" or "commonplace" and lacked its later negative connotations.)

How far can a scientific investigation be pursued, and what sort of knowledge can it yield? If we can only be said to have knowledge if that means knowledge in the rationalist sense, then it certainly looks likely that we can never attain knowledge. This is, of course, a profoundly sceptical conclusion. But what does it mean to talk about knowing things *in themselves*, as they actually are—in terms of, to use Locke's phrasing, their primary qualities? Is it not absurd to think that we can ever know things other than in terms of how they appear to us? Put differently, what is it that blocks our access to, and consequently our knowledge of, the very small; insufficiently powerful instruments or an inability to step outside of our sensory apparatus, our capacities to see, smell, touch, taste and hear? It is these questions that we will pursue next in the company of Locke's empiricist successors, Berkeley and Hume.

Chapter 10

Berkeley's Idealism

Berkeley's philosophy begins with the simple premise that what you see is what you get; or, in less vulgar terms, what you see is what there is. Following through what certainly looks like a straightforward idea results in a philosophical position that is disruptive and deeply counter-intuitive, but it is far from obvious that there is anything wrong with it. In Berkeley's hands what seem reasonable beginnings, when systematically thought through, generate apparently absurd conclusions. The challenge to the reader is to find a flaw in Berkeley's argument, or to show that there is some other reason to reject his approach. The alternative is to accept his claim that there is no matter (substance, in Lockean terms), and that there is only mind, ours and God's.

10.1 The denial of matter

Berkeley picks up first on Locke's claim that **substance**—things in terms of their **primary qualities**, bodies that are extended in space, that somehow underlie or support bodies as we experience them—in terms of their **secondary qualities**—is "only an uncertain supposition of we know not what." (*Essay*, I.IV.18) Berkeley's point is that you literally cannot make any sensible claims about something that cannot be sensed; if it cannot be known, how are we to say whether the claims made are true or false? How are we to evaluate them?

It seems obvious, though, that there are things as well as, in addition to, our perceptions of them. This is, as we have seen, the guiding thought at work in **representative realism** (7.2). None of us controls or directs the world as we perceive it, as we control and direct a daydream. The reality of an objective world that is independent of us seems to underpin the idea of a common world that we talk about and live our lives in. Denying the reality of this just seems absurd. But it is exactly here that Berkeley starts to turn the argument round, by arguing that believing in such an objective, independent material world leads to scepticism.

Berkeley fixes on a central doctrine in the physics of Descartes and Newton, that matter is inert and passive (barring gravity, which is deeply problematic). This is key to mathematical physics, that the motions of bodies, their behaviour on impact (how a car will crumple in an accident, for example), their mechanical properties (the load bearing capacity of a bridge, say), can be handled by mathematical methods (and gravity, of course, can be handled mathematically). If bodies are not so inert and passive, the world would be unpredictable; a bridge might suddenly detach itself at one end and roll up into a coil simply because it felt like it. This is the "disenchanting of nature", because it turns the world from a strange and mysterious place in which anything might happen to one in which things, at least, inanimate things, behave in entirely predictable law-like fashion.

The difficulty comes when the inertness and passivity of material things is put together with the idea that the mind is equally passive in perception. How can what is passive (material things) *cause* our experience of (ideas of) the world that we perceive? (You might recall that Descartes' proof of God in the Third Meditation (p.56, above) relies on the idea that a cause (God) must be greater than its effect (what there is, the Creation).) Put this way, what sense can it make to claim that what is passive, and hence cannot be a cause at all, can cause an effect—our perceptions of whatever-it-is?) This is Berkeley's statement of this problem in the first of the two works we will consider, his *Three Dialogues between Hylas and Philonous*:

Philonous Pray let me know what reasoning your belief of matter is grounded on; and what this matter is in your present sense of it.

Hylas I find myself affected with various ideas, whereof I know I am not the cause; neither are they the cause of themselves, or of one another, or capable of subsisting by themselves, as being altogether inactive, fleeting, dependent beings. They have therefore some cause distinct from me and them: of which I pretend to know no more, than that it is the cause of my ideas. And this thing, whatever it be, I call *matter*.

Philonous Tell me, Hylas, hath every one a liberty to change the current proper signification annexed to a common name in any language? For example, suppose a traveller should tell you, that in a certain country men might pass unhurt through the fire; and, upon explaining himself, you found he meant by the word *fire* that which others call *water*: or if he should assert there are trees which walk upon two legs, meaning men by the term *trees*. Would you think this reasonable?

Hylas No; I should think it very absurd. Common custom is the standard of propriety in language. And for any man to affect speaking improperly, is to pervert the use of speech, and can never serve to a better purpose, than to protract and multiply disputes where there really is no difference in opinion.

Philonous And doth not matter, in the common current acceptation of the word, signify an *extended*, *solid*, *movable*, *unthinking*, *inactive* **substance**?

Hylas It doth.

Philonous ...[S]hould [matter] be allowed to exist, yet how can that which is inactive be a cause; or that which is unthinking be a cause of thought? You may indeed, if you please, annex to the word *matter* a contrary meaning to what is vulgarly received; and tell me you understand by it an *unextended*, *thinking*, *active* being, which is the cause of our ideas. But what else is this, than to play with words, and run into that very fault you just now condemned with so much reason? I do by no means find fault with your reasoning, in that you collect a cause from the phenomena: but I deny that the cause deducible by reason can properly be termed *matter*. (*Three Dialogues*, p.101)

Berkeley's work is dense and often needs considerable unpacking. In his last sentence Philonous (i.e. Berkeley; "philonous" means *lover of mind*) agrees with Hylas ("hylas" means *matter*) that there is a cause of the phenomena (the phenomena being anything that we experience, as an occurrence or a fact), only he denies that matter—substance, things in terms of their primary qualities—can be the cause. This being the case we have to look elsewhere for the true cause. First, though, here is a passage in which Berkeley points out how a belief in substance, of a "something we know not what," leads to scepticism:

> As to ideas or unthinking things, our knowledge of these hath been very much obscured and confounded, and we have been led into very dangerous errors, by supposing a twofold existence of the objects of sense, the one intelligible, or in the mind, the other real and without the mind: whereby unthinking things are thought to have a natural subsistence of their own, distinct from being perceived by spirits. This ...is the very root of scepticism; for so long as men have thought that real things subsisted without the mind, and that their knowledge was only so far forth real as it was conformable to real things, it follows, they could not be certain that they had any real knowledge at all. For how can it be known, that the things which are perceived, are conformable to those which are not perceived, or exist without the mind?
>
> Colour, figure, motion, extension and the like, considered only as so many sensations in the mind, are perfectly known, there being nothing in them which is not perceived. But if they are looked on as notes or images, referred to things or archetypes existing without the mind, then are we involved all in scepticism. We see only the appearances, and not the real qualities of things. What may be the extension, figure, or motion of any thing really and absolutely, or in itself, it is impossible for us to know, but only the proportion or the relation they bear to our senses. Things remaining the same, our ideas vary, and which of them, or even whether any of them at all represent the true quality really existing in the thing, it is out of our reach to determine. So that, for aught we know, all we see, hear, and feel, may be only phantom and vain chimera, and not at all agree with the real things, existing in *rerum natura* [among the things of nature]. All this scepticism follows, from our supposing a difference between *things* and *ideas*, and that the former

> have a subsistence without the mind, or unperceived. It were easy to dilate on this subject, and show how the arguments urged by sceptics in all ages, depend on the supposition of external objects.
>
> So long as we attribute a real existence to unthinking things, distinct from their being perceived, it is not only impossible for us to know with evidence the nature of any real unthinking being, but even that it exists. Hence it is, that we see philosophers distrust their senses, and doubt of the existence of heaven and earth, of every thing they see or feel, even of their own bodies. And after all their labour and struggle of thought, they are forced to own, we cannot attain to any self-evident or demonstrative knowledge of the existence of sensible things. But all this doubtfulness, which so bewilders and confounds the mind, and makes philosophy ridiculous in the eyes of the world, vanishes, if we annex a meaning to our words, and do not amuse ourselves with the terms *absolute*, *external*, *exist*, and such like, signifying we know not what. I can as well doubt of my own being, as of the being of those things which I actually perceive by sense: it being a manifest contradiction, that any sensible object should be immediately perceived by sight or touch, and at the same time have no existence in Nature, since the very existence of an unthinking being consists in *being perceived.* (*Principles of Philosophy*, §§86–8)

The last paragraph is clearly an attack on Descartes, and Berkeley chooses his words with care. After all the struggle of thought, it is the lack of "any self-evident or demonstrative knowledge" of external things that Berkeley bemoans. This is a deliberate dig at **rationalism**, because if there are any innate ideas, they must be self-evident, because there cannot be any other kind of evidence for them. By "demonstrative" Berkeley means **deductive**, that is knowledge that can be demonstrated—proved or deduced, in more modern terms—from these self-evident first principles. So this is a carefully worded criticism of rationalism, but what it leaves standing is knowledge that is warranted in other ways (by experience and **induction**).

10.2 An idea can be like nothing but an idea

The evidence for such knowledge just is what we perceive; how could it be otherwise? But there is still the question of what is the *cause* of our experience, what there is that can be active in the requisite fashion. There are two parts to this, the first of which Berkeley slips in as if it is obvious:

> **Philonous** But how can that which is sensible be like that which is insensible? Can a real thing in itself invisible be like a colour; or a real thing which is not audible, be like a sound? In a word, can any thing be like a sensation or idea, but another sensation or idea? (*Three Dialogues*, p.92)

This is a puzzling claim. Of course we can tell a story about perception in terms of neurons and synapses and ganglions, of neurochemical transmitters and electrical activity in the brain, but how does this translate into my conscious experience of the world around me? And equally, as far as I can tell, into your similar conscious experience? The step at which brain activity, which can be observed by magnetic resonance imaging, turns into conscious experience, which is not obviously so detectable, is deeply mysterious. (MRI enables us to correlate brain activity and a subject's report of their thoughts, but it cannot explain one in terms of the other.) But what if we assume that there is only our own minds, that the only sort of "stuff" there is, is mind stuff? Then what seems mysterious—the means whereby physical/nervous activity is translated into consciousness—is demystified, because it is simply a transfer of mental content (ideas and sensations) from one (active) mind to another.

The second step concerns the cause of our experience, how it is that we perceive a common world:

> **Philonous** To me it is evident ...that sensible things cannot exist otherwise than in a mind or spirit. Whence I conclude, not that they have no real existence, but that seeing they depend not on my thought, and have an existence distinct from being perceived by me, there must be some other mind wherein they exist. As sure therefore as the sensible world really exists, so sure is there an infinite, omnipresent spirit who contains and supports it. (*Three Dialogues*, p.97)

What is objective is God, who is an active mind, constantly engaged in imprinting ideas on all the other minds that there are:

> Hence it is evident, that God is known as certainly and immediately as any other mind or spirit whatsoever, distinct from our selves. We may even assert, that the existence of God is far more evidently perceived than the existence of men; because the effects of nature are infinitely more numerous and considerable, than those ascribed to human agents. There is not any one mark that denotes a man, or effect produced by him, which doth more strongly evince the being of that spirit who is the *Author of Nature*. For it is evident that in affecting other persons, the will of man hath no other object, than barely the motion of the limbs of his body; but that such a motion should be attended by, or excite any idea in the mind of another, depends wholly on the will of the Creator. He alone it is who upholding all things by the Word of his Power, maintains that intercourse between spirits, whereby they are able to perceive the existence of each other. And yet this pure and clear light which enlightens everyone, is itself invisible. (*Principles of Philosophy*, §147)

Berkeley's philosophy is **idealist** because he holds that what we think of as real is actually ideal, in the sense that there are only ideas (there are minds as well, but these manifest themselves in terms of the ideas they contain). What there is, is mind stuff. But since, in Berkeley's terms, to be is to be perceived, there is nothing over and above mind stuff. This is the basis of Berkeley's so-called "master argument":

> For to what purpose is it to dilate on that which may be demonstrated with the utmost evidence in a line or two, to anyone that is capable of the least reflection? It is but looking into your own thoughts, and so trying whether you can conceive it possible for a sound, or figure, or motion, or colour, to exist without the mind, or unperceived. This easy trial may make you [i.e. the rest of us as upholders of the commonsense view that material things cause our perceptions] see, that what you contend for, is a downright contradiction. Insomuch that I am content to put the whole upon this issue; if you can but conceive it possible for one

extended moveable substance, or in general, for any one idea or any thing like an idea, to exist otherwise than in a mind perceiving it, I shall readily give up the cause: and as for all that *compages* of external bodies which you contend for, I shall grant you its existence, though you cannot even give me any reason why you believe it exists, or assign any use to it when it is supposed to exist. I say, the bare possibility of your opinion's being true, shall pass for an argument that it is so.

But say you, surely there is nothing easier than to imagine trees, for instance, in a park, or books existing in a closet, and nobody by to perceive them. I answer, you may so, there is no difficulty in it: but what is all this, I beseech you, more than framing in your mind certain ideas which you call *books* and *trees*, and at the same time omitting to frame the idea of any one that may perceive them? But do not you yourself perceive or think of them all the while? This therefore is nothing to the purpose: it only shows you have the power of imagining or forming ideas in your mind; but it doth not show that you can conceive it possible, the objects of your thought may exist without the mind: to make out this, it is necessary that you conceive them existing unconceived or unthought of, which is a manifest repugnancy [contradiction]. When we do our utmost to concede the existence of external bodies, we are all the while only contemplating our own ideas. But the mind taking no notice of itself, is deluded to think it can and doth conceive bodies existing unthought of or without the mind ...

It is very obvious, upon the least inquiry into our own thoughts, to know whether it be possible for us to understand what is meant, by the absolute existence of sensible objects in themselves, or without the mind. To me it is evident those words mark out either a direct contradiction, or else nothing at all. And to convince others of this, I know no readier or fairer way, than to entreat they would calmly attend to their own thoughts: and if by this attention, the emptiness or repugnancy of those expressions does appear, surely nothing more is requisite for their conviction. It is on this therefore that I insist, to wit, that the absolute existence of unthinking things are words without a meaning, or

> which include a contradiction. This is what I repeat and inculcate, and earnestly recommend to the attentive thoughts of the reader. (*Principles of Philosophy*, §§22–4)

To think what is thought of is to think it, so of course there cannot be anything that is not thought of. Does this prove that all there is, is mind stuff, or that to be is to be perceived? On the other hand, what does the assertion of matter as a something we know not what that underpins the world as we perceive it add to our experience?

It could be that Berkeley is right, but assuming that he is not, how can it be shown that he is wrong? The world as the Berkeleian experiences it is the same as the world as anyone else experiences it, after all. To find a way of rejecting the Berkeleian position one has first to establish what would count as disproof (or a proof, for that matter, if it is true). To borrow a piece of terminology from the philosophy of science what is needed is a "crucial experiment", one whose initial premises both parties can agree on, and one whose outcome will decide for one side or the other.

In looking for such an experiment, though, it may be helpful to pause and consider briefly Berkeley's position in terms of the theory of knowledge; as this is, after all, our main topic. To talk of knowledge one has to have a view on what knowledge is of. Berkeley argues that if there are only bodies, Cartesian/Lockean substances or material things, then we can never have knowledge of what there is, because what is inert and passive cannot be a cause of our sensations and ideas. We perceive things in terms of their secondary qualities, and these can only be in the mind (and not in the world).

The claim that there are only minds and only ideas (including sensations) accounts for our knowledge of bodies as we experience them (in terms of their secondary qualities). Our experience of "the world" around us is passive because its flip side is the active mind of God systematically imprinting ideas on us. It is because God is benevolent and does this systematically that the world is predictable, and this enables us to infer laws of nature (and more or less to predict how we, and other people, will behave in given circumstances). Further, Berkeley holds that my knowledge of my own mind is so immediate that it would be absurd even to question it, and even though I cannot perceive any other mind directly the evidence for the existence of other minds (particularly God's) is all around me. (It is tempting to say that it is self-evident that I have a mind, but since my mind is my self it can hardly be *evidence*, as a sticky floor is evidence that your housemate has spilt his orange juice. How can a mind have *evidence* of or for itself?)

This said, is there a flaw? His contemporaries thought his denial of matter absurd, and Samuel Johnson famously said "I refute him thus!" while painfully kicking a large stone. But of course Berkeley would simply reply that Johnson's sequence of ideas and sensations arose from God systematically imprinting ideas on his mind, so this is no refutation. But his contemporaries also questioned what scientists are looking into, if there are only ideas, and in particular what is going on when our perception of something changes when we see it through a microscope. If to be is to be perceived, how can a closer look at, say, the wing of an insect reveal an extraordinary tracery of structure? Again Berkeley can simply point to a systematic imprinting of ideas, but the sceptic may feel that this is an *ad hoc* clutching at straws. We are still at an impasse.

There is perhaps one way in which Berkeley's philosophy can be disproved; if an instance can be found in which how things seem to us is not how they are, and if this disparity persists after closer examination. If this can be shown then a gap has opened up between ideas and what they are of, and this can only be the case if there is something that is not an idea (i.e. that is not in, or of, any mind at all). But it is hard to think of any way in which this can occur. On the other hand, you may think that what explains everything in this fashion, and is immune to any objection, does indeed leave everything as it is—because it explains nothing.

Verificationism and falsificationism

Towards the end of the nineteenth century a new science of the mind began to emerge, pioneered by Freud. As we have seen (5.2), Descartes' approach to the mind does not entail that the mind is transparent to itself. Following the method of the *Meditations* enables the reader to uncover what they have in mind by their own efforts. Freud's approach is rather different, in no small part because he seeks to uncover very different mental phenomena than the truths of metaphysics sought by Descartes. Freud's individuals are driven by unconscious motives and desires, and uncovering these requires a new kind of mind mechanic or technician, the psychotherapist. Just as the scientist sets out to discover truths about the world by applying a body of laws and theories to disparate phenomena, so the Freudian analyst sets out to discover truths about the unconsious by studying disparate phenomena generated by the subject (word and picture associations, dreams, stream of consciousness diatribes, and so on).

The theories diverge in their notions of individual responsibility and accountability. A familiar pattern in the history of science is phenomena—theory—experiment, followed by evaluation. If the experiments work all is well, for the time being. But over time anomalies build up, generating pressure on existing theories, until a conceptual breakthrough occurs and a theory with greater explanatory power replaces the old. Freud's work coincides with the period during which Einstein developed relativity and a number of physicists developed quantum mechanics. The disparities between the rigorous, mathematical character of these theories and that of the new "sciences" of the mind bothered many people, among them the young Karl Popper.

Popper's central idea is that the mark of a scientific theory is that it is falsifiable (**falsificationism**), that there has to be a way in which it can be shown to be false. The traditional approach to scientific theories, dating back to Bacon's formulation of **induction** in the seventeenth century, is **verificationism**. The central claim of verificationism is that piling up positive instances makes a theory more secure. But Popper came to see that Freudian analysts can interpret almost any "evidence" their clients supply to substantiate their theories. Since the Freudian analyst is rummaging about in the subconscious, the part of the mind that is not consciously accessible by the subject themself, the analyst can claim direct knowledge of the contents of the subject's subconscious mind, whereas the subject themself can only come to know what is in their subconscious mind at second hand, as retold to them by their analyst. In all sorts of ways the autonomy and authority of the individual is undermined. The analyst claims knowledge on the basis of a body of theory, over-ruling whatever objections the subject may make, and if there is any contrary evidence, the analyst can postulate some deeper as yet undiscovered subconscious urge, the discovery of which will require a further course of analysis.

Falsificationism requires of a properly scientific theory that it indicates the circumstances in which the theory is disproved or falsified. Such a theory generates testable predictions so that if the predicted negative outcome comes about, the theory is shown to be false. Any theory that is unable to generate testable outcomes is not a properly scientific theory, and attempts to salvage it by *ad hoc* explaining aways of counter-examples is bad scientific methodology. Hence if any sequence of ideas can be explained away by reference to God, maybe Berkeley is doing theology and not philosophy? How can we tell?

Berkeley was aware of one possible set of counter-arguments to his approach. These come from Newtonian physics, from work which now falls outside the mainstream philosophical canon. Indeed, of his philosophical contemporaries, including Descartes and Locke, Berkeley wrote in his personal notebook "I see no wit [no great mental ability] in any of them but Newton, the rest are mere triflers." Berkeley's difficulty with Newtonian physics lies with the latter's arguments for absolute space, time and motion, independent of any thinking being.

Newton famously argues for absolute space, time and motion in the Scholium to Defn.8 in the *Principia.* But, as he admits, we cannot experience any of these, because all we can derive from the senses is relative; "instead of absolute places and motions we use relative ones, which is not inappropriate in ordinary human affairs, although in philosophy abstraction from the senses is required." (*Principia Mathematica*, p.411/66) But it is exactly such "abstracting from the senses" that Berkeley objects to, because all it leaves behind is a Lockean substance, a "something we know not what". Consequently Berkeley objects to Newton's work for the same reasons that he objects to Locke's, not least because a demonstration of absolute space, time or motion would serve to undermine Berkeley's own approach. It would count as proof of a phenomenon that is other than how it seems to us, whose appearance persists after closer examination.

In the Scholium to Defn.8 Newton puts forward an argument to demonstrate absolute motion, the "rotating bucket" argument. If you put a bucket full of water on a turntable and spin it at a constant angular velocity, the circular motion of the bucket is transferred to the water until they are stationary relative to one another. If you also stand on the turntable, the bucket and the water are stationary relative to you. Now imagine that the universe shares your motion, so everything looks to be at rest. How can you tell—how can you *know*—that you are moving? The answer, according to Newton, is right in front of you. The surface of the water is not flat, it is saucer-shaped, because the circular motion it experiences causes the water to recede from the axis of its circular motion. We have, it would seem, a proof of absolute motion, irrespective of any other motions the water/bucket/turntable/observer assemblage is subject to.

The agenda for the rotating bucket experiment becomes clear later in the Scholium. Descartes' **vortex theory** (8.1) has the planets carried round in their orbits by celestial vortexes. An observer on a planet in a Cartesian universe could argue that they are at rest, while all else is in (relative) motion; there would be no way of telling, because they are at rest within their vortex. The rotating bucket experiment shows that what appears to be at rest (the water relative to the bucket) is in fact in motion, shown by recession from the axis of circular motion.

Berkeley was well aware that a demonstration of absolute space, time or motion constitutes a counter-example to his approach. He specifically discusses the rotating bucket argument at *Principles* §114, but far from disproving the argument, just says "I do not see how this [the proof of absolute motion] follows from the experiment which was brought to prove it." The flaw in the rotating bucket experiment is that it privileges the location of the observer as the frame of reference for measurement, but this takes us too far afield. In 1915, when he was working on the theory of relativity, Einstein called his "happy thought" the idea that someone in freefall would not experience their own weight. They would still have the same mass, because they would still be the same amount of physical stuff, but they could have no perception of their own weight. If their bathroom scales, also in freefall, were attached to their feet, they would read zero. It seems to me as I sit here that I am static, although I know I am whirling through space at an astronomical speed. Maybe not everything is as it seems to us, but in realising this God has imprinted a new set of ideas on our minds; so, as it is not obvious that we are getting anywhere fast, we will move on to consider a trio of "isms".

10.3 Realism, idealism and phenomenalism

Realism, idealism and phenomenalism are three different ways of thinking about what there is, and about our relations to what there is.

Realism

Realism is the claim, more or less, that what there is exists independently of us, meaning that it will continue to exist whether or not we, or any other minded beings, are around to perceive it. On first acquaintance realism seems the obvious way to think about what there is; surely there is a world external to me, that existed more or less as it does now before I came along and will continue to exist, again more or less as it is now, when I am long gone? A now rather obscure German philosopher, J.G. Fichte, neatly characterises objects as "that which objects", literally, to what we try to do to it; objects persist, they are the stuff of the world that we try to shape to our ends. It is because we encounter objects that object in this way that we come to realise that we are agents, active in an objective world. If there were no "objects" we could never perceive ourselves as agents because there would be nothing, so to speak, to push against. Analogously an echo-sonar can be used to detect submarines or plot the contours of the sea-bed, by sending out a pulse that is reflected back.

Realism comes in two forms, **naïve realism** and **representative realism**. The former is naïve in the sense that it unproblematically takes what we perceive as a facsimile of what there is. But clearly we often make mistakes, and this casts doubt on naïve realism. So representative realism assumes that what we have in mind as a result of perception is a more or less accurate or more or less illusory representation of what there is. Instead of the things postulated by naïve realism (us, and what there is) representative realism postulates three sorts of things, what there is, ourselves, and mental representatives that act as intermediaries between us and the world. Locke's "ideas" are such intermediaries (7.2).

Representative realism has its own set of problems, with the nature of the relations that obtain in the two cases (between what there is and our ideas, and between our minds and our ideas). Intermediaries are helpful as a way of explaining illusion and error, and to maintain the independence of what there is from us, but they come at a price. As we have seen with both Descartes and Berkeley, postulating matter as inert (as the sort of stuff needed for mathematical physics) leads to certain difficulties. How are our ideas *caused* by what is external to us, if what is outside us is composed of causally inert substance? As we have seen in this chapter, thoughts like this this led Berkeley to **idealism**.

Idealism

Idealism is, broadly, the thesis that there exists nothing but minds and their ideas (their contents). As we have seen, with Berkeley, our ideas get into our minds because God systematically imprints them there. This gets over the problem of explaining how inert substances can be the cause of our ideas by abolishing the problem, but anyone with a robust sense of reality is unlikely to find Berkeley's approach satisfactory. Note that idealism does not entail denying that there is anything external to, and independent of, us. Rather it is denied that what is external to us is *material* as opposed to *ideal*. This is why Berkeley calls his philosophy "immaterialism".

A related approach is to think of a "total" mind, a composition of all consciousness with each individual mind comprising a minute part of this totality. This totality may be equated with God or it may include God as a part. But this tends towards mysticism and it is hard to see how such a view can be argued for, as opposed to simply asserted. Since theory of knowledge is about evidence and justification, we will fast-forward over this because the essential elements upon which to base a knowledge claim are lacking.

Phenomenalism

Phenomenalism is the view that there is no more to things than our perceptions of them. In a manner of speaking, after all, what more can there be? Phenomenalism is often encapsulated in Berkeley's famous phrase—"*esse* is *percipi*", to be is to be perceived. (*Principles of Philosophy*, §3) This is J.S. Mill's statement of phenomenalism:

> I see a piece of white paper on a table. I go into another room. If the phenomenon always followed me, or if, when it

> did not follow me, I believed it to disappear *è rerum naturâ* [from the things of nature], I should not believe it to be an external object. I should consider it as a phantom—a mere affection of my senses; I should not believe that there had been any Body there. But, though I have ceased to see it, I am persuaded that the paper is still there. I no longer have the sensations which it gave me; but I believe that when I again place myself in the circumstances in which I had those sensations, that is, when I go again into the room, I shall again have them ...The conception I form of the world existing at any moment, comprises, along with the sensations I am feeling, a countless variety of possibilities of sensation: namely, the whole of those which past observation tells me that I could, under any supposable circumstances, experience at this moment, together with an indefinite and illimitable multitude of others which though I do not know that I could, yet it is possible that I might, experience in circumstances not known to me. These various possibilities are the important thing to me in the world ...
>
> Matter, then, may be defined, a Permanent Possibility of Sensation. If I am asked, whether I believe in matter, I ask whether the questioner accepts this definition of it. If he does, I believe in matter: and so do all Berkeleians. In any other sense than this, I do not ...The reliance of mankind on the real existence of visible and tangible objects, means reliance on the reality and permanence of Possibilities of visual and tactile sensations, when no such sensations are actually experienced. (J.S. Mill, *An Examination of Sir William Hamilton's Philosophy*, p.179–80, 183)

Phenomenalism offers a solution to scepticism, but it comes at a price. It just seems fantastic to suggest that the contents of my fridge cease to exist when I shut the door. It is hard to see how to defend phenomenalism without making a move similar to Berkeley's, that is, relying on God to maintain whatever-it-is while no-one's looking.

Later recent attempts at rejecting phenomenalism rely on a more sophisticated set of arguments. If there a defining characteristic of twentieth century analytical philosophy, it is careful study of language and its uses as a way of getting at what we think and at what we know. I have tacitly deployed this approach in one or two places, here we will briefly acknowledge it and see how it works.

For phenomenalism to work, it has to be the case that everything that is said or thought about, say, the table in my dining room, can be expressed in sentences that state what a given individual will experience if they are in my dining room. It must be possible to do this in such a way that no appeal is made to the table itself, because (so the theory goes) there is no table over and above the possibility of sensations. The table is to be *reduced* to a set of such reports just as, for example, some early proponents of artificial intelligence thought that the human mind could be reduced to a set of instructions that can be coded into a computer.

In practice such a reduction is impossibly complex for limited beings with finite minds—for people like us. But this by itself is not a clear-cut principled refutation of phenomenalism. What it does make clear is that phenomenalism relies heavily on conditional statements of the form "*if* you are in my dining room, *then* you will experience dining table-like sensations," and this involves lots of less than trivial beliefs about law-like connections and regularities. If the criteria of a good theory include simplicity and plausibility, then phenomenalism scores rather badly and realism comparatively highly.

Chapter 11

Scepticism

Since Hume is often taken to be a sceptic—Kant says he is "perhaps the most ingenious of all sceptics, and is incontrovertibly the pre-eminent one with regard to the influence that the sceptical procedure can have on awakening a thorough examination of reason" (*First Critique*, A764/B792)—we should perhaps consider scepticism before turning to Hume. The topic of this chapter is what it is to know something-or-other in a general sense, rather than scepticism about the subject (ourselves) or the object (whatever it is) of knowledge.

Epistemology is traditionally defined as the study of what it is to know—to possess knowledge—irrespective of what the knowledge in question is of. But one can hardly treat of such matters without worrying about the *object* of knowledge, whether it is knowledge of ourselves or of the world around us that is at stake. Just as it is hard to see how anyone can learn to swim by lying on a table and going through the motions, so it is hard to see how one can study what it is to know without taking some account of the contents of one's purported episodes of knowing. In this respect the topic of scepticism is grist to the traditional epistemological mill, but I'll leave it as an open question whether abstract discussion of what it is to know is intelligible to anyone outside of academic philosophy.

11.1 Two kinds of scepticism

Scepticism traditionally comes in two flavours; *academic* or *dogmatic* scepticism, and *Pyrrhonian* or *mitigated* scepticism. The first is associated with the Academy in ancient Athens, the latter with Pyrrho (ca.365–ca.275BC). The distinctive characteristic of academic scepticism is a blanket denial that we can have knowledge at all. Pyrrhonian scepticism holds only that we cannot know whether or not we ever have knowledge. The first is dogmatic because it makes a definite claim, that we can never have knowledge. By contrast the second holds that the assertion that we cannot know is itself a claim to knowledge and as such is unwarranted. To begin with, though, we will lay down some ground rules as to what is, and what is not, interesting.

First of all let us say that if you know something, it is true. This does not mean that it is true because you know it, just that if it turns out to be false then all along you did not have knowledge. This is actually quite a strong condition. If you believe something, and it turns out to be false, then your belief was false, a situation most of us would accept with equanimity. By contrast to know something, and then find out that it is false—that you lacked knowledge all along—is a disturbing state of mind to be placed in. We feel secure in what we know, and finding out that we did not actually know all along is to feel betrayed and deceived.

Second, we need to consider the areas in which scepticism is of interest. If a question can be readily settled, by whatever means (visual inspection say, or googling it), or if it is too slight to be of much interest (whether it will be raining in the morning of 3 March next year, say), then in such cases sceptical doubt is of no great interest. Scepticism is of interest when we are faced with significant questions where it is far from obvious how we should go about solving them, or where there is no evident basis upon which to evaluate any suggested answer. Whether we have free will or whether everything that happens is pre-ordained, whether there are laws of nature or whether it is just cosmic chance that regular and apparently predictable sequences of events occur, or whether the world around us looks the same to me as it does to you, for example. It is not obvious how we should go about answering such questions and what criteria we should use to judge any proposed answer. We seem to be, as Kant might put it, nudging up against the boundaries of reason.

In this section I will state briefly these two forms of scepticism and then consider how they fit with other themes discussed here. Near the beginning of Berkeley's *Three Dialogues* Hylas accuses

Philonous of being a sceptic, because the latter asserts that there is no such thing as matter or substance (Philonous, as we saw in the last chapter, asserts that there are only minds and their ideas). The dialogue continues:

Philonous Pray, Hylas, what do you mean by a sceptic?
Hylas I mean what all men mean, one that doubts of everything. (*Three Dialogues*, p.60)

It is not clear so far which version of scepticism Hylas has in mind, so Philonous goes on thus:

Philonous He then who entertains no doubt concerning some particular point, with regard to that point cannot be thought a sceptic.
Hylas I agree with you.
Philonous Whether doth doubting consist in embracing the affirmative or negative side of a question?
Hylas In neither; for whoever understands English, cannot but know that doubting signifies a suspense between both.
Philonous He then that denieth any point, can no more be said to doubt of it, than he who affirmeth it with the same degree of assurance.
Hylas True.
Philonous And consequently, for such his denial is no more to be esteemed a sceptic than the other.
Hylas I acknowledge it.
Philonous How cometh it to pass then, Hylas, that you pronounce me a sceptic, because I deny what you affirm, to wit, the existence of matter? Since, for aught you can tell, I am as peremptory in my denial, as you in your affirmation. (*Three Dialogues*, p.61)

Philonous and Hylas are arguing over academic and Pyrrhonian scepticism, where the Pyrrhonian sceptic suggests (asserts would be a word the Pyrrhonian would temperamentally decline) that for such questions, suspending one's belief or disbelief is the appropriate response. If Philonous's views are Berkeley's, then Berkeley is not a sceptic in either sense, because he thinks he can prove the existence of God and that there is no such thing as matter or substance. Whether his arguments are good ones is a topic we have already considered, but what we have not considered is whether these claims go beyond what we are rationally entitled to; that is, whether Berkeley makes claims that go beyond the bounds of reason into a realm of fanciful speculation. One way of reading Hume, as we will see in the next chapter, is as an attempt to bring Berkeley's flights of fancy down to earth.

A second question is this. If we have only the evidence of our senses, does a refusal to go beyond the evidence amount to scepticism? The basis of the mathematical physics that Descartes thought should come after the *Meditations*, realised in Newton's *Principia Mathematica*, lies in observing and measuring regularities and then finding mathematical equations and expressions for them. If the mathematical models work, if they track the regularities, then arguably there is no point in speculating about hidden mechanisms, because such speculation amounts to little more than flights of fancy. This question matters for interpreting Hume, because there is a long-running dispute whether Hume is a sceptic (and if so, in what sense), or whether in sticking to the evidence that is available to us and limiting the conclusions he is prepared to draw, he is simply applying scientific method to the study of minded beings—people like us. It is this question, in particular, that we will pursue in the next chapter.

Before this, as philosophy is an interactive discipline, here are some questions to ponder: can you draw a clear distinction between academic and Pyrrhonian scepticism? How would you defend the one against the other? How can the Pyrrhonian sceptic defuse (reject, after all, would be too strong a term) the academic sceptic's claim that denying that we can know one way or the other is itself ultimately a dogmatic claim? How can the academic sceptic refute the allegation that denying we can know is itself a claim to knowledge? And finally; how should the academic and the Pyrrhonian sceptic manifest their beliefs in their day-to-day actions?

11.2 Global scepticism

Here is a question that examiners are rather fond of; is it possible to doubt all your beliefs? In a paper on truth Russell considers a man who thinks he could be mistaken about (could doubt) all his beliefs:

> If he is mistaken in this belief, then *all* his beliefs are mistaken, which is what he is believing; therefore he is not mistaken; therefore he is right in believing that *all* his beliefs are mistaken, and therefore this belief is mistaken. (*On the Nature of Truth*, p.46)

Global scepticism would seem to be absurd. The claim that nothing is immune from doubt is self-defeating if the claim that nothing is immune from doubt is itself doubtful. But what if the would-be global sceptic accepts this and simply says that there is no single belief that we can be sure of; that maybe we do indeed have knowledge but cannot know that it is knowledge? This approach seems altogether stronger and more difficult to combat.

Global scepticism is challenged by the Cambridge philosopher G.E. Moore in a paper published in 1939 called "Proof of an External World." Moore puts forward a number of propositions that, he says, are so self-evident that there is no sense in which a rational being can coherently and intelligibly deny them:

> I can prove now, for instance, that two human hands exist. How? By holding up my two hands, and saying, as I make a certain gesture with the right hand, "here is one hand", and adding, as I make a certain gesture with the left, "and here is another". And if, by doing this, I have proved *ipso facto* the existence of external things, you will all see that I can also do it now in numbers of other ways: there is no need to multiply examples. (*Proof of An External World*, p.165–6)

By way of example, Moore later gave these:

> I am at present, as you can all see, in a room and not in the open air; I am standing up, and not either sitting or lying down; I have clothes on, and am not absolutely naked; I am speaking in a fairly loud voice, and am not either singing or whispering or keeping quite silent; I have in my hand some sheets of paper with writing on them; there are a good many other people in the same room in which I am; and there are windows in that wall and a door in this one. (*Certainty*, p.171)

Moore's point is that anyone who tries to deny the truth of these claims has forfeited any right to be taken seriously as a rational being. It is not that they are intelligibly going wrong, such that we can point out the error of their ways. Rather their behaviour is simply irrational and unintelligible, such that we reach for the straightjacket and the chemical cosh. Nevertheless it is not obviously impossible to give counter-examples to Moore's examples of the blindingly obvious. I may be dreaming (**dreaming argument**) or it could all be an illusion, I could be trapped in a *Matrix*-like world in which I am "really" a **brain in a vat** systematically fed electric pulses that cause "me" to think I am this person living this life, thinking these thoughts and writing these words. How can one respond to such challenges?

There are two possibilities. One is the "counterfeit" argument; you cannot have counterfeit (fake) coins or bank notes in circulation unless real ones exist, otherwise what would they be counterfeit copies of? Similarly you can't have doubts unless they ensue on prior beliefs. After all, how could we have the concept of doubt, if we don't first have the concept of belief? And must we not have held beliefs before we arrived at this concept of belief? Could there be a concept of colour in a world consisting entirely of blind people? I'm not saying this is conclusive, because it might still be conceivable that we could doubt all our beliefs one by one. I'm not at all sure, though, that even this is intelligible. Can I doubt that I am a thinking thing who is thinking that maybe everything I think is false? Perhaps there is something to be said for the "defence of scepticism" reading of the *Meditations* after all!

The second, and rather more prosaic, approach is to lay a trap for the global sceptic. When they say, "I doubt everything!" throw a can of baked beans at them and watch them catch it. Then ask them if they truly doubt the existence of an external world. Please, though, do not try this one at home.

We are now in a position to see why scepticism, which seemed so challenging in antiquity and in the seventeenth and eighteenth centuries, has now largely lost its teeth. Scepticism is primarily a philosopher's concern and it thrives in environments in which enquiry is conceptual rather than concrete or experimental. As we saw in our brief look at Aristotle's physics (6.2), pure conceptual enquiries are prone to pure conceptual rebuttals; one can invent a Cartesian romance to fit what little evidence there is. Once enquiries shift into a more scientific vein, scepticism based on conceptual objections can start to look silly. Particularly when such theories as relativity and quantum mechanics knock out arguments based on extending our commonsense intuitions about space, time, waves and particles.

On the other hand if someone tells you that machines will one day be more intelligent than people, that human activity does not affect the climate, or that the next round of political reforms will cause measurable improvements in health or education or whatever the flavour of the month is, a robust sceptical approach is appropriate because there are good grounds to show that they do not know what they are talking about.

Chapter 12

Hume's "Experimental Method"

Philosophers are fond of saying that we do not do "isms", that "isms" are a Bad Thing. But we all have a craving for classifying and ordering as a way of making things familiar and understandable, as a way of getting a handle on what would otherwise be unmanageable. So looking for patterns and continuity is a valuable exercise as long as we do not end up losing sight of what is interesting about a square peg by hammering it too vigorously into a round hole.

The labelling of Hume as a sceptic dates back to Kant. At the end of his monumental *Critique of Pure Reason* (1781) Kant names Hume as the quintessential sceptic, and this label has tended to stick ever since. Unfortunately Kant based his opinion on Hume's *Enquiry* rather than his *Treatise*, as the latter was not available to him, so we will begin with the *Treatise* and suspend judgment on the question of Hume's commitment to scepticism.

12.1 The experimental method

The full title of Hume's *Treatise* is "A Treatise of Human Nature, Being An Attempt to Introduce the Experimental Method of Reasoning into Moral Subjects", where by a "moral subject" Hume does not mean a subject like history or economics, he means a person capable of reasoning about how to act in any given circumstances, and then capable of acting accordingly. A moral subject

A

TREATISE

OF

Human Nature:

BEING

An ATTEMPT to introduce the experimental Method of Reaſoning

INTO

MORAL SUBJECTS.

Rara temporum felicitas, ubi ſentire, quæ velis; & quæ ſentias, dicere licet. TACIT.

VOL. I.

OF THE

UNDERSTANDING.

LONDON:

Printed for JOHN NOON, at the *White-Hart*, near *Mercer's-Chapel*, in *Cheapſide*.

MDCCXXXIX.

Facsimile of the title page of Hume's *Treatise*.

differs from a Cartesian extended body which, you will recall, is inert and incapable of thought or self-generated action. So Hume tells us that his plan is to apply scientific method—the experimental method—to reasoning about moral subjects—people, rational animals—as opposed to applying scientific method to reasoning about rocks and planets and the physical/chemical make-up of the things around us. We would expect, then, to see careful observation of the phenomena coupled with a story that draws back from speculating about inner mechanisms and hidden powers. Whether or not this is scepticism is, as mentioned in the last chapter, a question that, for the moment, we will keep an open mind about.

Given that Hume's plan is to apply the experimental method we would expect him to take as his raw material what it is readily observable. This is how Hume begins the *Treatise*:

> All the perceptions of the human mind resolve themselves into two distinct kinds, which I shall call *impressions* and *ideas*. The difference betwixt these consists in the degrees of force and liveliness, with which they strike upon the mind, and make their way into our thought or consciousness. Those perceptions, which enter with the most force and violence, we may name *impressions*; and under this name I comprehend all our sensations, passions and emotions, as they make their first appearance in the soul. By *ideas* I mean the faint images of these in thinking and reasoning; such as, for instance, are all the perceptions excited by the present discourse, excepting only, those which arise from the sight and touch, and excepting the immediate pleasure or uneasiness it may occasion. I believe it will not be very necessary to employ many words in explaining this distinction. Everyone of himself will readily perceive the difference betwixt feeling and thinking. (*Treatise*, p.1–2)

Hume goes on to distinguish the simple and the complex:

> There is another division of our perceptions, which it will be convenient to observe, and which extends itself both to our impressions and ideas. This division is into *simple* and *complex*. Simple perceptions or impressions and ideas are such as admit of no distinction nor separation. The complex are the contrary to these and may be distinguished into parts.

> Though a particular colour, taste, and smell are qualities all united together in this apple, it is easy to perceive they are not the same, but are at least distinguishable from each other. (*Treatise*, p.2)

He then considers the relation of ideas to impressions:

> The first circumstance, that strikes my eye, is the great resemblance betwixt our impressions and ideas in every other particular, except their degree of force and vivacity. The one seem to be in a manner the reflection of the other; so that all the perceptions of the mind are double, and appear both as impressions and ideas. (*Treatise*, p.2–3)

So far, then, he has established one general proposition:

> That all our simple ideas in their first appearance are derived from simple impressions, which are correspondent to them, and which they exactly represent. (*Treatise*, p.4)

It is, as Hume says, "the full examination of this question [that] is the subject of the present treatise," and it is indeed from this basis, the data of the mind, that Hume spins out the story he has to tell about ourselves as rational agents engaged with the world around us.

Here is an experiment to try out at home. Sit back in your chair and look at what you see in front of you. Now try and drain what you see of all conceptual content; what is left? It is tempting to say something like patches of colour at varying distances from me, but brief reflection suggests that distance is not something we actually see. If you think of yourself as a camera what you have is a flat two-dimensional representation. We add in perspective because we know that things are at different distances from us—but at the moment it is what we *know* that we seek an origin or basis for, so we need to bracket out everything that we bring to bear in interpreting our perceptions.

What remains is simple impressions or, rather, our ideas of simple impressions. This is the raw material that we derive from sensation but, so far, it does not *mean* anything. This leads to a second stage in which Hume considers the relations that obtain between simple ideas. The idea again is to observe the data and not smuggle in illicit meanings or interpretations. Hume proposes seven kinds of relations:

1. The first is *resemblance*: and this is a relation, without which no philosophical relation can exist; since no objects will admit of comparison, but what have some degree of resemblance. But though resemblance be necessary to all philosophical relation, it does not follow, that it always produces a connection or association of ideas. When a quality becomes very general, and is common to a great many individuals, it leads not the mind directly to any one of them; but by presenting at once too great a choice, does thereby prevent the imagination from fixing on any single object.

2. *Identity* may be esteemed a second species of relation. This relation I here consider as applied in its strictest sense to constant and unchangeable objects ...Of all relations the most universal is that of identity, being common to every being, whose existence has any duration.

3. After identity the most universal and comprehensive relations are those of *space* and *time*, which are the sources of an infinite number of comparisons, such as distant, contiguous, above, below, before, after, etc.

4. All those objects, which admit of *quantity*, or *number*, may be compared in that particular; which is another very fertile source of relation.

5. When any two objects possess the same *quality* in common, the *degrees*, in which they possess it, form a fifth species of relation. Thus of two objects, which are both heavy, the one may be either of greater, or less weight than the other. Two colours, that are of the same kind, may yet be of different shades, and in that respect to admit of comparison.

6. The relation of ***contrariety*** may at first sight be regarded as an exception to the rule, that no relation of any kind can subsist without some degree of resemblance. But let us consider, that no two ideas are in themselves contrary, except those of existence and non-existence, which are plainly resembling, as implying both of them an idea of the object; though the latter excludes the object from all times and places, in which it is supposed not to exist.

7. All other objects, such as fire and water, heat, and cold, are only found to be contrary from experience, and from the contrariety of their *causes* or *effects*; which relation of cause and effect is a seventh philosophical relation as well as a natural one. (*Treatise*, p.14–5)

We will allow evidence from all five senses, along with ideas derived from reflection on sensation (that I feel bodily sensations, am aware of correspondences between ideas derived from sight and ideas derived from touch, and so on). What is not allowed, because there is no basis for it, is any more to such relations than our perception of them. If I see that *these* look like *those*, that some are here and some over there, there are more of them than something else and they're somewhat darker in colour, and so on, these are relations that I perceive directly. There is no more to them than what can be seen. There is no explanation or reason why things are such-and-such rather than thus-and-so, they just are as they are and that is how we perceive them to be.

It would seem to be essential to human beings that we are beings with plans, projects and intentions. We set out to achieve these and in the process of so doing we reason about the world around us. In short, while we are interested in how things happen to be related now, we are also interested in bringing it about that things are related differently in the future; perhaps that there will be a house where I am standing now, or that I'll be in Edinburgh next week. We are interested in changing the relations that obtain between things so we are interested in finding out what changes are possible and how to bring them about.

On the basis of what we know so far there is no obvious way in which we can see how to bring changes about. Further, we want our actions to have predictable and systematic consequences. If I set out to drive to Edinburgh I do not want my car to change into a pumpkin north of Carlisle or find that the M1 has suddenly looped back on itself. I want things to be reliable, to be **projectible** into the future. This brings us to far and away the most important and influential aspect of Hume's philosophy, his views on cause and effect.

12.2 Cause and effect

The essence of **projectibility** is that causal relations are the only relations that take us beyond the evidence that is presented to us by the senses alone. That something looks like or is the same as something else, is next to or greater in size or number or darker than something else, is derivable solely from sensation. (So far relations of time have been skipped over, because succession is an aspect of cause and effect.) In general, though, Hume's point is that only causal relations extend our knowledge beyond what can be derived from sensation alone. Assuming that we have an idea of cause and effect, and given the self-imposed austerity of Hume's approach, where did we get this idea from? From which impressions is the idea derived?

Hume's approach is to reflect on sequences of events that we are tempted to think of as causal sequences. If something happens, say a flood of water suddenly falls on your head, you expect there to be a cause of your drenched state. You would probably look for this cause; perhaps a pipe has been leaking in the room above and the weight of water has finally brought the ceiling down. There is a reason for what happened, and when you call the insurance assessor you can give them an explanation. That is, the concepts of *cause*, *reason*, and *explanation* are very close together, because (that word again!) they are the concepts that we use to make sense of the world around us. The challenge for Hume's project, for the "experimental method of reasoning", is to derive an account of these concepts—in particular that of cause and effect—from the phenomena themselves, without invoking some mysterious psychological process (an innate idea, perhaps) as an explanatory *Deus ex machina*.

Given that there are causal sequences, that we know something in such cases that is not obviously a feature of the sequences themselves, what is it that we know and where did this knowledge originate? Hume soon finds on thinking about sequences of events that what they have in common is spatial location, that something impinges on something else, as the impact of a hammer drives a nail into a wall (waiving the case of "action at a distance"—gravity—as it is problematic in any case), and that causes precede their effects in time. The former Hume calls *contiguity*, the latter *succession*. But these by themselves are not sufficient to identify a sequence as causal. If you put the bin out on a Thursday morning it does not cause the rubbish collectors to empty it in the way that flicking a light switch causes a light to go on. Of course there is a relation in both cases and a story to be told, but there also seems to be a difference. However, as we will soon see, for Hume both of these are instances of a straightforward causal connection, by the standard of the third, defining, aspect of the causal relation, *constant conjunction*:

> It is easy to observe, that in tracing this relation, the inference we draw from cause to effect, is not derived merely from a survey of these particular objects, and from such a penetration into their **essences** as may discover the dependence of the one upon the other. There is no object, which implies the existence of any other if we consider these objects in themselves, and never look beyond the ideas which we form of them. Such an inference would amount to knowledge, and would imply the absolute contradiction and impossibility of conceiving any thing different. But as all distinct ideas are separable, it is evident there can be no impossibility of that kind. When we pass from a present impression to the idea of any object, we might possibly have separated the idea from the impression, and have substituted any other idea in its room.
>
> It is therefore by *experience* only, that we can infer the existence of one object from that of another. The nature of experience is this. We remember to have had frequent instances of the existence of one species of objects; and also remember, that the individuals of another species of objects have always attended them, and have existed in a regular order of contiguity and succession with regard to them. Thus we remember to have seen that species of object we call *flame*, and to have felt that species of sensation we call *heat*. We likewise call to mind their constant conjunction in all past instances. Without any further ceremony, we call the one *cause* and the other *effect*, and infer the existence of the one from that of the other. In all those instances, from which we learn the conjunction of particular causes and effects, both the causes and effects have been perceived by the senses, and are remembered: but in all cases, wherein

> we reason concerning them, there is only one perceived or remembered, and the other is supplied in conformity to our past experience.
>
> Thus in advancing we have insensibly discovered a new relation betwixt cause and effect, when we least expected it, and were entirely employed on another subject. This relation is their *constant conjunction.* Contiguity and succession are not sufficient to make us pronounce any two objects to be cause and effect, unless we perceive, that these two relations are preserved in several instances. (*Treatise*, p.86–7)

It is from the observation of regularity that we come up with the idea of cause and effect. It is important to note that Hume is not denying that there may be an explanation of such sequences of events if we could penetrate to the essences of things. (Compare Berkeley's remarks near the end of 10.1, and Locke's views on real essences at 7.5.) The fact is, though, that we cannot or, at least, could not in Hume's day, so what remains is the impressions from which we derive our ideas, and it is from impressions of contiguity, succession, and constant conjunction that we derive our idea of cause and effect. (The conceptual puzzles at the heart of quantum mechanics and statistical mechanics suggest we are little better off in the early twenty-first century.) We see such sequences of events occurring around us and from them we derive laws, and these laws enable us to predict, so far as it is in our power, what will happen, and to adjust our actions accordingly to achieve what we set out to achieve.

At this point Hume certainly seems to have achieved what he set out to achieve, namely, to derive an account of our mental abilities from what is available to us in sensation alone. That is, he has shown how our ideas can be traced back to impressions that derive from sensation alone, including reflection on such ideas (as it is from such reflection that the idea of constant conjunction arises), and how we can systematise them (in terms of relations) and project them into the future (in terms of cause and effect).

In the examples above, it was suggested that there is a sense in which putting the bin out does not cause it to be collected, whereas hammering a nail does cause it to go into a wall. In the first case most of us would revolt against calling this a causal relation as we are dealing with agents who possess freewill. It is a striking consequence of Hume's theory that this distinction is largely erased:

> A prisoner, who has neither money nor interest, discovers the impossibility of his escape, as well from the obstinacy of the jailer, as from the walls and bars with which he is surrounded; and in all attempts for his freedom chooses rather to work upon the stone and iron of the one, than upon the inflexible nature of the other. (*Treatise*, p.406)

Why is a tunnel more likely to get you out of jail? Hume's argument is to do with regularity and projectibility, and the truth is that people are very largely predictable. The interesting question is, then, why would most people bridle at the suggestion that they are broadly as predictable as a toaster or a microwave? Hume's account implies that psychological laws are on a par with physical laws, that there is no difference in principle between the chemistry of the inert gases and treating victims of post-traumatic stress disorder. Lower levels of success in the latter are a function of the complexity of the systems involved, rather than anything to do with freewill or some purported categorical difference between things and minded beings. If this does not seem right to you, can you think of an argument to show that Hume is wrong?

One response might be to suggest that psychological laws, like the laws of economics, are inherently statistical in a way that physical laws are not. At sea level ice always melts at 0°C whereas some people will always buck the trend, whatever the trend may be. But in fact some scientific laws are irreducibly statistical. The second law of thermodynamics says that heat will flow from a hotter body to a cooler one, but given the case of two gases some particles will buck the trend (some cooler particles will travel to the hotter one).

12.3 Hume as sceptic?

Before turning to Kant, who certainly did think that Hume's approach is flawed, we need to consider briefly the claim that Hume is a sceptic. In large part this is an historical claim. Hume's *Treatise* fell, as he put it, dead-born from the press, failing even to agitate religious zealots whom, Hume felt sure, would be sufficiently irritated to bring the book to the attention of a wider public. Realising that part of the reason for the *Treatise*'s poor reception lay in its presentation, Hume recast many of its arguments in his first *Enquiry*, published in 1748, nine years after the first two books of the *Treatise*.

The *Enquiry* is more polished than the *Treatise*, but philosophically less interesting. Looking for a vehicle to propagate his ideas led Hume to scepticism and, as great writers often do, he used it as a way of presenting his argument; much as Descartes used his *Meditations* as a vehicle to get over his views on physics, as we saw in 5.1. Of course this is a controversial suggestion, but sometimes you have to do whatever it takes to get your ideas across. (A prominent American computer scientist once remarked that he never worries about people stealing his ideas, as normally he has to ram them down their throats to get them accepted. This is an interesting observation on the sociology of knowledge.) This is also a curious argument to put forward because it is a meta-philosophical rather than a straightforwardly philosophical argument. It is not an argument on the same footing because it is not a point-by-point attempt at a refutation as, for example, Berkeley's rejection of Locke's views on primary and secondary qualities is a point-by-point rejection.

What it illustrates is a form of philosophical argument that can look like a change of subject rather than a refutation. Was Hume a sceptic? Arguably not, but it does not follow that he is a dogmatist (if being dogmatic is the opposite of being sceptical). Rather I would argue that Hume used scepticism as a convenient vehicle, while at bottom adhering to Newton's "hypotheses non fingo"; I do not feign hypotheses, I do not speculate when the evidence runs out. In Hume's own words:

> Nothing is more requisite for a true philosopher, than to restrain the intemperate desire of searching into causes, and having established any doctrine upon a sufficient number of experiments, rest contented with that, when he sees a further examination would lead him into obscure and uncertain speculations. In that case his inquiry would be much better employed in examining the effects than the causes of his principle. (*Treatise*, p.13)

In a similar vein, in the *Enquiry* he says this:

> It is confessed, that the utmost effort of human reason is to reduce the principles, productive of natural phenomena, to a greater simplicity, and to resolve the many particular effects into a few general causes, by means of reasonings from analogy, experience, and observation. But as to the causes of these general causes, we should in vain attempt their discovery; nor shall we ever be able to satisfy ourselves, by any particular explication of them. These ultimate springs and principles are totally shut up from human curiosity and inquiry. Elasticity, gravity, cohesion of parts, communication of motion by impulse; these are probably the ultimate causes and principles which we shall ever discover in nature; and we may esteem ourselves sufficiently happy, if, by accurate inquiry and reasoning, we can trace up the particular phenomena to, or near to, these general principles. The most perfect philosophy of the natural kind only staves off our ignorance a little longer: as perhaps the most perfect philosophy of the moral or metaphysical kind serves only to discover larger portions of it. Thus the observation of human blindness and weakness is the result of all philosophy, and meets us at every turn, in spite of our endeavours to elude or avoid it. (*Enquiry*, p.30–1)

Is this scepticism, or is this the careful scientist paying due heed to the available evidence and to the conclusions that can be drawn from this evidence alone? Or, put differently, would we want to say that scientific method—the "method of . . .experiments and observations", as Cotes puts it in a thinly-veiled dig at Descartes (6.2)—is inherently sceptical? Whatever the case, reading Hume famously woke Kant from his "dogmatic slumbers", as we will see in the next chapter.

Chapter 13

Kant's "Copernican Revolution"

Kant is the last of the great philosophers we will look at individually, as after his time the landscape changes somewhat. A look at Kant first will help to firm up some of the issues concerning *a priori* and *a posteriori* knowledge, scepticism, and our knowledge of the external world.

13.1 Necessity and applicability

The question we have been working with throughout is how it is that we seem to know more than we can straightforwardly get out of experience. As we saw in Chapter 2, on the *Meno*, our knowledge of geometrical/mathematical truths seems to go far beyond what we can get out of experience alone. We can see that any number of plane triangles have internal angles that add up to 180°, but there is no obvious way to substantiate the claim that all plane triangles at any point in the past, present and future have internal angles that add up to 180° (plane triangles because the internal angles of triangles on convex and concave surfaces do not add up to 180°). Numbers and sets (a set is any collection of things of whatever kind thought of as a single entity) and other mathematical objects similarly seem to retain their properties irrespective of their spatio-temporal location, and to persist as they are whether or not they are ever conceived of or perceived by rational beings.

It also seems to be the case that mathematical truths are **necessary**. It is coherent to deny that the sun will rise tomorrow, but it is not coherent to claim that $2 + 2 = 5$ or that you can draw a plane triangle with angles of 90°, 90°, and 90°. There is no obvious way of conceiving of a world in which $2 + 2 = 5$. Imagine a world in which, for example, rulers are made of rubber, and give unpredictable and erratic results. How could we build roads and houses in such a world? If someone claims repeatedly that $2+2=5$ we would not think they have made an intelligible mistake that they can be talked out of. Rather we would think that they are not in their right mind.

Together with this quality of necessity goes that of universal **applicability**. Everything that there is can be counted—can be numbered—and its position relative to other things can be specified (for example, 3 metres to the left of whatever it is, at an angle of 47° to the back wall). In this sense mathematics is literally everywhere even if its objects—numbers, sets, geometrical objects in terms of their Platonic forms—are nowhere and nowhen.

We have, then, three characteristics of geometrical and mathematical truths:

- not derivable from experience alone,
- necessary,
- universally applicable.

Turning to Newton's laws of motion, as we saw in 6.1, they equally do not seem to be derivable from experience alone. They also seem to be necessary truths in the sense that if we came across circumstances in which one of the laws appears not to hold, we assume that there is some unidentified causal factor at work. We ascribe necessity to these laws because we hold them to be immune from revision, and because we cannot conceive of circumstances in which we would consider revising them. (You might like to consider whether this sort of necessity is the same as that claimed for $2 + 2 = 4$, earlier. Is there a difference between "we cannot conceive of *this* ($2+2=4$) being false" and "we cannot conceive of circumstances in which we would be prepared to give up *this* (that a body continues in a state of rest or uniform motion unless acted on by an external force)"? Commonly the former is called **logical necessity**, the latter **physical necessity**. How would you state the difference—if there is a difference—between them?)

Unsurprisingly, Newton's laws of motion certainly seem to be universally applicable. Thus we find that at least some laws of nature share the three characteristics of geometrical/mathematical truths listed above, of not being derivable from experience alone, of being necessary, and of being universally applicable. There seems to be a harmony these laws of nature and the truths of geometry and mathematics, and this apparent harmony stands in need of explanation.

Socrates, as we saw in 2.1, sought to account for our knowledge of geometrical and mathematical truths by a "glorious truth" of reincarnation coupled with a doctrine of learning as recollection. Descartes opts for a programme of intellectual exercise or meditation in which we establish such truths as what is clear and distinct, self-evident to the attentive mind. Newton sticks to what is mathematically demonstrable and eschews further speculation as ungrounded hypotheses. Locke and Berkeley are chary of going beyond what we can perceive, while Hume says that the only way in which we can extend our knowledge beyond what we can perceive—by means of causal relations—is a matter of mental habituation arising from our observations of contiguity, temporal succession and constant conjunction between our ideas.

Kant rejected all of these, and in particular regarded Hume's assimilation of cause and effect, and consequently the laws of nature, to a mental habit no different in principle to always remembering to leave one's keys in a pot by the door, as unacceptably sceptical. (As we saw in the previous chapter Kant based his opinion of Hume on the *Enquiry* rather than the *Treatise*.) His solution is a brilliantly simple strategy of turning the question around. Given these necessary and universally applicable truths, and given that they cannot be derived from experience alone, perhaps it is the case that they derive from us; could it be that they are brought to bear by us in experience and provide the framework that we use to order and structure our experience into a systematic and intelligible whole? If this is the case then we can account for the fact that they are:

- not derivable from experience alone
  - Yes, because they are prior to experience (are *a priori*) and are brought to bear *in* experience,
- necessary
  - Yes, because they are the means whereby we conceive

of the world, so it is literally inconceivable that they could be otherwise,

- universally applicable
  - Yes, because again they are deployed by us in experiencing what there is; they are like a pair of tinted spectacles that we cannot take off (or, better, that we are not even aware of wearing).

Much as Copernicus proposed a radical shift in our thinking about the universe, that it is centred on the sun rather than on the earth, so Kant proposed a radical shift in the subject-matter of philosophy, from what there is to the mental apparatus that we bring to bear in making sense of our world:

> I should think that the examples of mathematics and natural science, which have become what they are now through a [Newtonian/Galilean] revolution . . . were remarkable enough that we might reflect on the essential element in the change in the ways of thinking that has been so advantageous to them, and, at least as an experiment, imitate it insofar as their analogy with metaphysics, as rational cognition, might permit. Up to now it has been assumed that all our cognition must conform to the objects; but all attempts to find out something about them *a priori* through concepts that would extend our cognition have, on this presupposition, come to nothing. Hence let us once try whether we do not get farther with the problems of metaphysics by assuming that the objects must conform to our cognition, which would agree better with the requested possibility of an *a priori* cognition of them, which is to establish something about objects before they are given to us. This would be just like the first thoughts of Copernicus, who, when he did not make good progress in the explanation of the celestial motions if he assumed that the entire celestial host revolves around the observer, tried to see if he might not have greater success if he made the observer revolve and left the stars at rest. (*First Critique*, B xv–xvi)

So far this is somewhat programmatic, albeit suggestive. Kant next associates arithmetic with our experience of time, and geometry with our experience of space. Arithmetic goes together with time because Kant thinks that our grasp of number arises from a capacity to count, and in counting we, so to speak, number the seconds. Geometry goes together with space because arguably we experience things primarily as coloured shapes that stand in spatial relations to one another. By "geometry" Kant means Euclidean geometry, as contained in the thirteen books of Euclid's *Elements*. Because space and time are bound up with relations in which things or events stand to one another, they are not things that we perceive directly. The mathematical/geometrical is thus neatly put together with experience in a broad sense; whatever we experience has spatio-temporal location (relative to us) prior to any question of what sort of thing it is. This is Kant's statement of his position with respect to geometry:

> Pure mathematics, and especially pure geometry, can have objective reality only under the single condition that it refers merely to objects of the senses, with regard to which objects, however, the principle remains fixed, that our sensory representation is by no means a representation of things in themselves [noumena], but only of the way in which they appear to us [phenomena]. From this it follows, not at all that the propositions of geometry are determinations of a mere figment of a poetic fantasy, and therefore could not with certainty be referred to actual objects, but rather, that they are valid necessarily for space and consequently for everything that may be found in space, because space is nothing other than the form of all outer appearances, under which alone objects of the senses can be given to us. Sensibility, whose form lies at the foundation of geometry, is that upon which the possibility of outer appearances rests; these, therefore, can never contain anything other than what geometry prescribes to them. (*Prolegomena*, 4:p.287)

While this gives us objects of perception that are spatially and temporally located, so far our knowledge of things doesn't obviously go beyond that of a camera that keeps a record of when it is used and of its location and direction. Our experience as agents in the world goes beyond this, because we use it as a basis for making judgments about how the world is, about how we might want it to be different and how we should set about making it different. Given that we have such knowledge, the question is, what must be the case—what sort of conceptual apparatus must we possess—for such knowledge to be possible for creatures like us?

Many of the judgments we make about our world are based on knowledge derived from experience. The beliefs we form in this way are, on the whole, inductive generalisations, and consequently they are **defeasible**; we would be prepared to revise them if the evidence warranted revision (the discovery of black swans in Australia, for example). Another way of putting this is to say that such truths are **contingent**, meaning that they are dependent on ways in which the world happens to be. A contingent truth is one that could be other than it is, a truth that could conceivably be otherwise. It is also a truth that can be rationally or coherently denied, as it is perfectly rational or coherent to deny that the sun will rise tomorrow. Such truths are *a posteriori*, not *a priori*, so it is not these truths that are of present concern. The truths that are of interest are those that are *a priori*; are necessary, universally applicable, and not obviously derivable from experience.

We have seen, briefly, that Kant's next move is to put arithmetic and geometry together with time and space. His idea is that Newton's laws of motion and, hopefully, other fundamental laws of nature, of physics and chemistry, will follow from a concentrated study of the workings of our mental apparatus. Kant calls this part of science "pure natural science", as opposed to empirical generalisations (otherwise known as rules of thumb) derived directly from experience. In line with his "Copernican revolution" Kant claims that whatever is *a priori* in these laws must come from us, that it must in some sense originate in our mental apparatus. Schematically there is what we perceive, consisting of objects situated relative to us in space and time. In perception the mind is essentially passive. Then there is judgement, in which we judge what sort of thing something is, and what it can and cannot do in terms of the *a priori* concepts we have in mind (if it is a heavy body it will fall if unsupported, for example). In making judgements the mind is active. Kant holds that such judgements are objective because any minded being should make the same judgement because, so he holds, we are all born with the same mental apparatus insofar as we are dealing with what is (what can be known) *a priori*. Finally there are subjective judgements such as "this tastes good to me" or "this seems heavy to me", which relate to my own personal experience and may or may not be shared by others. This is how Kant words it:

> We will therefore have to analyse experience in general, in order to see what is contained in this product of the senses and the understanding, and how the judgment of experience is itself possible. At bottom lies the intuition of which I am conscious, i.e., perception, which belongs solely to the senses. But, secondly, judging (which is attributed solely to the understanding) also belongs here ...
>
> The given intuition [perception] must be subsumed under a concept, which determines the form of judging in general with respect to the intuition [perception], connects the empirical consciousness of the latter in a consciousness in general, and thereby furnishes empirical judgements with universal validity; a concept of this kind is a pure *a priori* concept of the understanding, which does nothing but simply determine for an intuition [perception] the mode in general in which it can serve for judging.[1] The concept of cause being such a concept, it therefore determines the intuition [perception] which is subsumed under it, e.g., that of air, with respect to judging in general—namely, so that the concept of air serves, with respect to expansion, in the relation of the antecedent to the consequent in a hypothetical judgment. The concept of cause is therefore a pure concept of the understanding, which is completely distinct from all possible perception, and serves only, with respect to judging in general, to determine that representation which is contained under it and so to make possible a universally valid judgment.
>
> Now before a judgment of experience can arise from a judgment of perception, it is first required: that the perception be subsumed under a concept of the understanding of this kind; e.g., the air belongs under the concept of cause, which determines the judgment about the air as hypothetical with respect to expansion. This expansion is represented not as

[1] [The concept of *cause* is the most familiar example of such a pure *a priori* concept, a concept that we use to connect a subject and a predicate in a judgement (for example, the wetness of the pavement was caused by last night's rain). The full list is at *Prolegomena*, §21.]

> belonging merely to my perception of the air in my state of perception or in several of my states or in the states of others, but as *necessarily* belonging to it, and the judgment: the air is elastic, becomes universally valid and thereby for the first time a judgment of experience, because certain judgments occur beforehand, which subsume the intuition [perception] of the air under the concept of cause and effect, and thereby determine the perceptions not merely with respect to each other in my subject, but with respect to the form of judging in general (here, the hypothetical), and in this way make the empirical judgement universally valid.
>
> If one analyses all of one's synthetic judgments insofar as they are objectively valid, one finds that they never consist in mere intuition [perceptions] that have, as is commonly thought, merely been connected in a judgment through comparison, but rather that they would not be possible if, even beyond the concepts drawn from intuition [perception], a pure concept of the understanding had not been added under which these concepts had been subsumed and in this way first combined into an objectively valid judgment. (*Prolegomena*, 4:p.300–1)

Kant claims that if we analyse a judgement, we see that it contains an *a priori* part contributed by the mind, consisting of a spatio/temporal framework and a number of *a priori* concepts (e.g. cause, universal or particular, possibility, existence, necessity), and an *a posteriori* part contributed in perception. Thus "concepts drawn from [perception]" are *a posteriori* ones, that is, ones that do not arise solely from the mind but are based on experience, for example, colours, classifications of plants and animals, sounds, tastes, and so on.

This separation of the constituents of judgement is a crucial step because Kant's aim is to assimilate the most general laws of nature to our understanding, to the *a priori* part of the rational faculty. As we have seen, Kant assumes that we are all capable of making such judgements. (I should also say, incidentally, that these two passages from Kant's *Prolegomena* are far the hardest quoted in the present book, but there is no substitute for the original texts and I hope you will get the general drift of Kant's argument from them.) Anyone not capable of making such judgments would count as irrational, not as somehow mistaken. The fact that we are all beings whose minds embed an understanding that includes the laws of nature is what makes these laws necessary and universally applicable, and the reason they cannot be readily derived from experience is that they are already there in experience. Hence we deduct from what we have in mind what we get in experience (sensation), and are left with the contents of the rational faculty, the framework of space and time and *a priori* concepts that we bring to bear in making sense of the world around us. Hence we get what is *a priori*, that is, literally, what is prior to experience.

This is why, for Kant, Newton's extraordinary advances in physics arose from a combination of mathematical acumen and thought experiments. On this picture radical scientific innovation does not come from experiments in the usual sense. It comes from gifted individuals thinking long and hard about the contents of their own minds. The greatest and most important knowledge comes not from examining and studying the world around us but from, to use Hegel's phrase, "thought thinking about itself". Certainly one might wonder how much of scientific theory derives from such reflection, and how much derives from reflection on the contents of a mind well stocked with existing theories and experimental results. The highly mathematical nature of modern science makes such questions extremely difficult to address.

But this leaves a strange and impenetrable secret at the heart of Kant's approach, the realm of things-in-themselves, called by Kant the *noumenal*, as opposed to the *phenomenal* world that we experience (as in the passage at p.160, above). Because we can only perceive things as they are ordered by us in space and time, with space and time themselves our contribution, and because judgement, whether subjective or objective, inescapably involves a conceptual contribution from us, there remains an unknowable world somewhere beyond the bounds of our perceptual and conceptual apparatus. This is rather disturbing.

Chapter 14

Science and Epistemology

In this chapter first we will bring our story more or less up-to-date, and second we will round out some issues concerning the concepts of knowledge and justification; that is, belief, knowledge, evidence and degrees of justification, reliabilism, coherentism and foundationalism, and the tripartite definition of knowledge (knowledge as justified true belief).

14.1 An uneasy relationship

In the last chapter we looked at Kant's views on our knowledge of the *a priori*; space and time, as applications of the concepts of geometry and number that we bring to bear in experience, and certain high-level concepts—principally that of cause and effect—that are not derived from experience but are brought to bear by us in making sense of our experience.

This approach goes together with the idea that philosophy is prior to natural science, that thinking about thought itself, isolating what is *a priori*, is the first source of knowledge. On this view philosophy generates the conceptual scheme used by other disciplines, so natural science has to wait on and follow on from the discoveries of philosophers. The last great proponent of this view was G.W.F. Hegel (1770–1831), whose three volume *Encyclopaedia of the Philosophical Sciences*, devoted to logic (the study of forms of thinking, the *a priori*), and its applications (the philosophy of nature and the philosophy of mind), was intended to provide a comprehensive philosophical foundation for all subsidiary disciplines, including natural science. But even in Hegel's lifetime such an approach was increasingly untenable and as the nineteenth and the twentieth centuries progressed the star of natural science rose, eclipsing that of philosophy. So in this section I will briefly bring our story up-to-date as a backdrop to the topics we will look at in the rest of this chapter. First, we will look at two major developments in the nineteenth century that undermined Kant's conception of *a priori* knowledge, concerned with geometry and number.

Geometry

Kant's argument, as we have seen, is that we perceive space in terms of Euclidean geometry because Euclidean geometry is, so to speak, hard-wired into our minds. This is the basis of Kant's account of the necessity and universal applicability of geometry; these follow automatically from the idea that in making sense of the spatial arrangements of things in the world around us we bring Euclidean geometry to bear.

In the nineteenth century, though, non-Euclidean geometries were developed, principally by Riemann and Lobachevsky. Imagine you are standing at the North Pole. You walk due south until you get to the equator—about 6200 miles—and turn left through 90°, and walk a further 6200 miles. Now turn left again through 90° and walk due North, again about 6200 miles, until you are back at the North Pole. Your original path is 90° to you, so you have walked three straight lines, come back to the point you started at—just as if you had walked a triangle—yet the included angles add up to 270°. How can this be? The reason is that the earth is not flat, it is a sphere, so applying geometry on the surface of the earth over any significant distance involves non-Euclidean geometry (geometries in which the included angles of a triangle add up to more than, or less, than 180°, on surfaces that are either convex or concave).

A similar effect occurs with parallel lines. Euclid's geometry includes the *parallel postulate*, that given a line L and a point x not on L, exactly one line can be drawn through x that never meets L, however much we extend the two lines. But this does not work on a curved surface. Two lines either converge, in which case they touch sooner or later, or they diverge, in which case they may eventually loop back on one another.

The crunch comes with Einstein's theory of relativity. This shows that space is not conceptually distinct from time, and that space-time is curved (5.3). But this does not prove that Kant was wrong. It could be that we can only perceive space in terms of Euclidean geometry because while the range of our perceptual apparatus, unaided, is enormous—we can, after all, "see" stars light years away (we see the light emitted from them years ago)—our capacity to discriminate is not nearly so great. To "see" the red shift effect we have to use sophisticated experimental techniques and apply sophisticated theoretical methods. Perhaps, then, we perceive what there is in terms of Euclidean geometry, and apply theoretical techniques to make appropriate adjustments. Nevertheless we have some reason to be wary of making *a priori* claims about Euclidean geometry.

Further, this suggests that the Kantian project of identifying what is necessary and *a priori* as aspects of judgements contributed by us is possibly misguided. Kant devised this approach as a bulwark against Humean scepticism, which he associates with empiricism. If non-Euclidean geometry furnishes grounds for rejecting Kant's approach, and if pragmatically the success of modern science undermines the grounds for scepticism, then perhaps we can reasonably adopt empiricism. This question will be taken up later in the present chapter.

Number

As we also saw, Kant associates number with time, with our ability to count; as someone might count under their breath 1, 2, 3, and so on, so Kant suggests that we are aware of number. Many events might happen in the time it takes me to count to one thousand, and these events will all have a temporal ordering, an ordering I can ascribe numbers to.

The development that matters here in the nineteenth century was that of axiomatisation. In the case of number the axioms are those set out by the Italian mathematician Guiseppe Peano in 1899:

1. 0 is a number.
2. The immediate successor of a number is a number.
3. 0 is not the immediate successor of any number.
4. No two numbers have the same immediate successor.
5. Any property belonging to 0, and also to the immediate successor of every number that has the property, belongs to all numbers (principle of mathematical induction).

Instead of thinking of numbers as things, axiomatisation enables us to think of them in terms of a set of rules. These rules define the permissible manipulations of symbols, that is, they define not *things* but *structures*. A line of clothes pegs on a washing line is as good a model of the beginning sequence of the natural numbers as saying one, two, three, and so on, out loud, and does not require any reference to a thinking being. On this view mathematics is about structures, not things, and mathematical knowledge is knowledge of rules for manipulating these structures, not knowledge of things. For example:

A two-legged duck has two legs.

is a tautology, because the subject ("a two-legged duck") already contains the predicate ("has two legs"). Similarly:

$$2 + 2 = 4$$

is a tautology, as a meaning of "4" is "2+2" (other "meanings" are 2^2, $5 - 1$, and so on). In a post office you might find a letter-sorting machine; you put a pile of letters into the hopper, it scans the postcodes, and outputs the letters into a series of pigeon holes. Barring malfunctions the output is the same as the input, only it is sorted (a structure has been imposed upon it). The underlying idea is that arithmetic, and mathematics generally, is a colossal collection of tautologies that adds no content; it is a collection of techniques for ordering and arranging whatever content it is applied to.

As with non-Euclidean geometry, we have grounds for discarding Kant's approach. The necessity of mathematics does not arise from its realisation in the rational faculty, imposed by us in making judgements about the world. Rather it reflects the idea of mathematics as a contentless collection of structures, a set of techniques that acts as a bridge between sets of propositions; from an initial plan to build a bridge over a river, perhaps, to a finished scheme fully calculated and optimised to cope with weather conditions and traffic loads.

Furthermore these developments, coupled with the fallout from the development of quantum mechanics in the 1920s, undermined the idea of theory of knowledge as a subject distinct from natural science, let alone prior to it. If theory of knowledge is about how we form accurate, reliable beliefs about what there is—beliefs that are appropriately justified, are true—as science (I'll drop the "natural" from hereon) is the most successful enterprise going in this area, perhaps theory of knowledge should be refocussed as a descriptive study of scientists about their work.

Reliabilism

Here is an example of this sort of approach. If a dark speck moves close enough to a frog, it will shoot out its tongue, catch it and eat it. A frog has to eat, so frogs that survive will be those that are most effective at discriminating and catching passing insects. If a frog catches an insect, its belief-forming mechanism ("that's edible and within range") has generated the answer "true". If it fails (it was not an insect but a speck of soot, it was not in range) then it's belief-forming mechanism has generated an error (a falsehood). A frog that is consistently wrong will not live for long and probably will not reproduce, so over time it is likely that frogs will manifest more accurate belief-forming mechanisms.

Successful frogs will be those with a reliable belief-forming process when it comes to food. This is known as **reliabilism**, and it applies in similar ways to people. Those who are effective in achieving what they set out to do are likely to be more successful, all else being equal. Their beliefs are justified because they are generated by reliable belief-forming mechanisms. So if we want to study knowledge, perhaps we should study the belief-forming processes of successful people. In this way the theory of knowledge would become a branch of psychology, in which we study successful people, in order to extrapolate principles that underlie their success and to see how we can teach these principles to others. We would, of course, also look at the less successful as an example of how not to go about things.

While reliabilism has some attractive and plausible aspects, it does not obviously have much to say as to *why* a belief-forming process is reliable. A clairvoyant might be reliable but there is no obvious reason why they are reliable. Someone who is a reliable guide to wine, for example, will be someone who has studied their subject, has a wide experience of tasting and has a sophisticated palette. You would have good reason to trust such a person's judgement in their field, to think that their judgments are likely to be accurate and consistent. But this seems to sever the link between knowledge and evidence, because if somebody is reliable then they just are reliable; that they are reliable is its own justification. This is why reliabilism is called a "non-justificatory" approach in theory of knowledge. We will look more closely at such relationships between concepts in 14.5.

14.2 Verificationism and evidence

Before the emergence of epistemology as a descriptive enterprise, it was profoundly influenced in the 1920s by *logical empiricism* (the label "logical empiricism" is now more generally used than "logical positivism"). The most influential proponents of this approach comprised the *Vienna Circle*, a group mainly of scientists with philosophical interests. Logical empiricism combined the approach to the *a priori* sketched above, that the *a priori* is a collection of tautologous principles that can be used for ordering or structuring data given in experience, that is, what is given *a posteriori*. Combining the latter with the former yields, so they argued, properly scientific knowledge about the world and about ourselves (the task of a properly scientific psychology).

Logical empiricism begins with the "Verification Principle" (**verificationism**). This is Ayer's version:

> We say that a sentence is factually significant to any given person, if, and only if, he knows how to verify the proposition which it purports to express—that is, if he knows what observations would lead him, under certain conditions, to accept the proposition as being true, or to reject it as being false. (*Language, Truth and Logic*, p.16)

The connection of truth-value with observation is significant. By **proposition** is meant, more and less, what a sentence says:

the sentences "Snow is white" and "La neige est blanche" say *that snow is white* so, since we are interested in the things themselves, it is propositions we are interested in. Here is a programmatic statement of logical empiricism's central doctrines from Rudolf Carnap, one of the leaders of the Vienna Circle:

> What, then, is the method of verification of a proposition? Here we have to distinguish between two kinds of verification: direct and indirect. If the question is about a proposition which asserts something about a present perception, e.g. "now I see a red square on a blue ground", then the proposition can be tested directly by my present perception. If at present I do see a red square on a blue ground, the proposition is directly verified by this seeing; if I do not see that, it is disproved. To be sure, there are still some serious problems in connection with direct verification. We will however not touch on them here, but give our attention to the question of *indirect* verification, which is more important for our purposes. A proposition P which is not directly verifiable can only be verified by direct verification of propositions deduced from P together with other already verified propositions.
>
> Let us take the proposition P_1: "This key is made of iron." There are many ways of verifying this proposition; e.g.: I place the key near a magnet; then I perceive that the key is attracted. Here the deduction is made in this way:
>
> *Premises*:
>
> P_1 "This key is made of iron"; the proposition to be examined.
> P_2 "If an iron thing is placed near a magnet, it is attracted"; this is a physical law, already verified.
> P_3 "This object—a bar—is a magnet"; proposition already verified.
> P_4 "The key is placed near the bar"; this is now directly verified by our observation.
>
> From these four premises we can deduce the conclusion:
>
> $\mathbf{P}_5$ "The key will now be attracted by the bar."

This proposition is a prediction which can be examined by observation. If we look, we either observe the attraction or we do not. In the first case we have found a positive instance, an instance of verification of the proposition P_1 under consideration; in the second case we have a negative instance, an instance of disproof of P_1.

In the first case the examination of the proposition P_1 is not finished. We may repeat the examination by means of a magnet, i.e. we may deduce other propositions similar to P_5 by the help of the same or similar premises as before. After that, or instead of that, we may make an examination by electrical tests, or by mechanical, chemical, or optical tests, etc. If in these further investigations all instances turn out to be positive, the certainty of the proposition P_1 gradually grows. We may soon come to a degree of certainty sufficient for all practical purposes, but *absolute* certainty we can never attain. The number of instances deducible from P_1 by the help of other propositions already verified or directly verifiable is *infinite*. Therefore there is always a possibility of finding in the future a negative instance, however small its probability may be. Thus the proposition P_1 *can never be completely verified*. For this reason it is called an **hypothesis**.

So far we have considered an individual proposition concerning one single thing. If we take a general proposition concerning all things or events at whatever time and place, a so-called natural *law*, it is still clearer that the number of instances is infinite and so the proposition is an hypothesis.

Every assertion P in the wide field of science has this character, that it either asserts something about present perceptions or other experiences, and therefore is verifiable by them, or that propositions about future perceptions are deducible from P together with some other already verified propositions. If a scientist should venture to make an assertion from which no perceptive propositions could be deduced, what should we say to that? Suppose, e.g., he asserts that there is not only a gravitational field having an effect on bodies according to the known laws of gravitation, but also a *levitational field*, and on being asked what sort of

> effect this levitational field has, according to his theory, the answer is that there is no observable effect; in other words, he confesses his inability to give rules according to which we could deduce perceptive propositions from his assertion. In that case our reply is: your assertion is no assertion at all; it does not speak about anything; it is nothing but a series of empty words; it is simply without sense. (Carnap, *Philosophy and Logical Syntax*, Kegan Paul Trench Trubner & Co., 1935, p.10–14)

This is a **foundationalist** approach to knowledge, because it begins with evidence derived from observation and builds on it (using tautologous principles). Logical positivism is, as noted earlier, now called "logical empiricism", because it is a combination of logic (the tautologous ordering/structuring principles we met earlier) and **empiricism** about what is observable, what can be experienced. This is why Carnap denies that any law derived from experience in this way—by induction—can ever be *certain*. Hence Carnap denies, in the spirit of Locke (7.5), and Hume (12.2), that necessary truths can ever be derived from experience alone.

At the end of Chapter 12 we touched on the question of whether scientific method is inherently sceptical. We can now refine this and suggest that we regard the results of scientific investigations as inherently **defeasible**; we are prepared to accept that any scientific theory we hold at present may be revised, and may even be proven false, at some point in the future. If induction and defeasibility are central to scientific methodology, then applying them to scientific enquiries themselves shows that theories held at past times have been revised and/or superseded, suggesting that our present theories may in turn be revised and/or superseded by new theories as yet unthought. It is of course possible that some of our present theories will never be revised or superseded; does that make them *necessary* truths, true everywhere and everywhen? How can we ever know? With this background in place, we will turn to perhaps the biggest debate in twentieth century epistemology, that between **foundationalism** and **coherentism**.

14.3 Foundationalism and coherentism

Foundationalist approaches invariably begin with a basic kind of knowledge that is derived directly from experience. All our complex knowledge, our theories about what there is and how it behaves, is then constructed out of this basic knowledge. In Carnap's version basic knowledge is expressed in "protocol sentences", of the form "red here now", or perhaps "at 10:07 am the needle on the dial pointed to seven." These are supposed to be direct, unmediated reports of what is given in perception. This is akin to phenomenalism, as 10.3; what is given is what an observer would experience if they were appropriately situated.

Our scientific theories about the world are then constructed out of basic knowledge using mathematical/logical structures, as discussed in 14.1. These do not add any content, because the content is all derived from experience. They provide a means of ordering our experience and, by **induction**, arriving at generalisations that have the character of laws. We use these laws to predict what will happen, and to enable us to bring about events that we want to bring about. Any higher-order sentence, in any language, can be translated into a (probably extremely long) protocol sentence, and it can then be seen (by getting oneself into an appropriate set of circumstances) whether or not it obtains, whether it is true or false. If it cannot be so translated then it is fiction or poetry and certainly not something that involves knowledge (it is, in Carnap's words, "simply without sense"). (Waiving, for the sake of the argument, sentences about inaccessible regions of space-time, the undocumented past and infinitely complex problems, as 3.2.)

The strongest argument for foundationalism is the "infinite regress" argument. If a belief is justified there must be something that justifies it. If this is in turn a belief, then it must itself be justified, and so on, until we get to basic beliefs that are justified in experience. There must be such basic beliefs, so the foundationalist claims, or there can be no knowledge.

But inevitably, perhaps, this approach runs into problems. Of all intellectual disciplines science prides itself on objectivity, that its results are independent of any given individual, are testable and repeatable. An experiment to demonstrate nuclear fusion, for example, should be repeatable in any laboratory with suitable equipment and should yield the same results. But protocol sentences are essentially private, they are my report of what I am perceiving now. What if I am colourblind, or if there is some other reason why my senses are askew; is my report more or less correct than anybody else's? Perhaps, in the end, we need an "ideal observer", someone whose perceptual apparatus is, well, *normal.* Whatever they would report in such circumstances is therefore correct, only it seems rather absurd to suggest that such an observer is to be present at any experiment to note the results. Objectivity, after all, is supposed to be an aspect of the experiment, not of the observer.

A related difficulty concerns the idea that there is a "given", that somehow what there is is such that if we could bracket all our preconceptions we would, in any given set of circumstances, have essentially the same perceptions. In 5.4, I suggested that after reading the passages from Descartes' with the "defence of Cartesian physics" reading in mind, that you tried reading them with the "response to scepticism" reading in mind. The idea is to show that something can seem different if you approach it with a different set of preconceptions, even though you know it is the same thing. There is a whole series of optical "illusions" designed to demonstrate this:

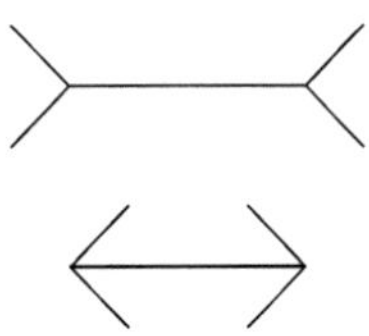

Which line is longer?

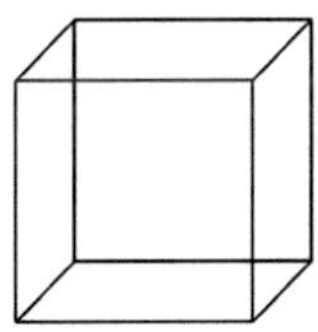

The Necker cube

The duck-rabbit

With the lines on the left, even though you *know* they are the same length, it is hard not to think that the line with the arrows pointing out is longer. In this case you almost have to override the evidence of the senses. The second is the Necker cube; does it go into or come out of the page? The third is the duck-rabbit figure; it can be seen as a duck or as a rabbit but not, it would seem, as both at the same time. But which is it? A duck, or a rabbit? Or, put differently, in these three cases, what is *given* in perception? Surely there should be an answer here.

The implication is that there is no clear-cut "given" in perception, because somehow our experience is enmeshed with our perceptions—with what we expect to see—all the way down (to use a foundationalist metaphor). Even if we do not always see what we expect to see, what we see is, it would seem, coloured by our conceptual scheme, our general "take" on the world. Our ideal observer needs not only to have a "normalised" sensory apparatus, but also to have a known, calibrated conceptual scheme so we can be sure that they are perceiving aright, and this seems a tad absurd. Put differently, although it is reasonable to claim that our perceptions are *caused* by things around us, there is no clear-cut causal relation between what there is and what we perceive.

Given these difficulties there was a split among the logical empiricists between supporters of **foundationalist** approaches and supporters of **coherentist** approaches to knowledge and justification. The distinctive coherentist claim is that a belief can only be justified by a belief, that our body of knowledge about the world is a systematic, interlocking body of theory that is mutually self-supporting. This is akin to Berkeley's claim, at 10.2, that an idea can only be caused by an idea; that what is in the mind can only get there through the influence of another mind. Coherentists argue that since there is no adequate way of explaining how material things are supposed to cause our knowledge of the world, maybe we can justify our beliefs on the grounds that they fit into an integrated, interlocking body of theory about what there is, where this body of theory generates successful prediction.

It is not hard to see why, historically, coherentism has been associated with **rationalism** and **idealism**, and foundationalism with **empiricism**. An extreme rationalist, holding that knowledge is of Platonic forms, is unlikely to have much time for the evidence of the senses (c.f. 4.2). An idealist, holding that what there is consists of minds and ideas, is also unlikely to pay much attention to a material world of substantial things. By contrast, foundationalism is grist to the empiricist's mill. It seems rather odd, then, to find a move towards coherentism occurring among a group of philosopher-scientists in the 1930s.

The reasons, in the main, have to do with problems of objectivity and the difficulty of isolating a suitable "given", something that everybody can perceive in given circumstances and agree on. Once it is accepted that there is no clear way of agreeing on a world given as it is in perception, coupled with the fact that our theories are generally effective in enabling us to get around the world and achieve what we set out to achieve, the coherentist would seem to have a reasonable basis to argue that "experience" as such has evaporated. If the idea of being massively or consistently wrong drops out, then the way to justify a belief is not to ignore the experience but just say, it fits with my other beliefs and it works; what more is there to say? Perhaps we should even embrace Kant's phenomenal/noumenal divide (13.1), ascribing our coherent body of theory to the phenomenal and letting the noumenal take care of itself.

At first, foundationalism seems more attractive and intuitively more plausible. If you look at the world around you, what do you see? Would a bushman from the Kalahari see your computer mouse? According to legend when the Spanish invaded some parts of South America the natives were astonished when men got off their horses; what strange beasts, that they can suddenly divide in two like that! But on the other hand coherentism seems to ride roughshod over our intuitions that our knowledge is based on experience. Coherence theory can look like a "consistent fairy story", plausible, no doubt, but possibly entirely wrong as, perhaps, Communism is as a theory of human nature.

14.4 Epistemology naturalised

Foundationalism and coherentism embody two different ways of constructing a systematic body of theory about what there is, theories that encapsulate our knowledge of the world. They share a common characteristic, both invoking an intellectual standpoint external to a body of knowledge. **Foundationalism** says, more or less, begin with a set of basic experiences that are warranted in some suitable way, and then construct a theory—a body of knowledge—on this basis in accord with suitably transparent principles. **Coherentism** says, more or less, take all your beliefs and render them consistent, by suitable means, and in this way attain a suitably comprehensive body of knowledge that enables you to get around in the world and do whatever it is you set out to do. But what of the status of these theories of knowledge themselves? Do they constitute further knowledge, or knowledge of some other kind? Are they *knowledge* at all?

If epistemology is itself an autonomous intellectual discipline, with its own subject matter and its own body of knowledge, then it seems there must be something or other that it is knowledge *of*. In a way I have tried systematically to evade this problem by talking about knowledge in the context of science, but now we need to address this head-on and ask, is theory of knowledge itself a discipline with its own subject matter? Or will we find on closer inspection that we have been chasing a will-o'-the-wisp, that intellectual disciplines, particularly science, can get along well enough without any "help" from epistemology?

Epistemological theories characteristically tell a story about us and our ways of finding out about the world. If you put a small child in a room full of toys, she will find her way around and interact with the toys. Over time she learns about shapes and colours and textures, about what you can and what you cannot do with any given toy. Given time she learns where the fridge and the phone are, about clothes and makeup and boys and so on and so forth. Small children are budding epistemologists, gradually developing a theory about the world around them, a theory that enables them to progressively navigate their world and warp it to their ends.

One way of thinking about epistemological theories is as "rational reconstructions" of the processes the child goes through. It is not obvious that by observation we can tell whether a particular child is a rationalist or an empiricist, a foundationalist or a coherentist, but hopefully we can devise a rational reconstruction that enables us to account for their observed behaviour on the basis of their prior interactions and experiences and a plausible theory of what they could deduce or infer from their experiences.

The difficulty with "rational reconstruction" is that it seems to demand some way of characterising experience independently of the theory—the body of knowledge—itself. Given the child in the room full of toys, we have to be able to say things like "She has realised that the square blocks go through the square holes in the box" and "She has worked it out that if she stands on the igloo she can reach the jelly on the table," but to do this we attribute to her a capacity to perceive things as they are. But we know from the Necker cube and the duck-rabbit that there is no clear-cut "as they are" that we see things in terms of. This implies that epistemology is not separate from the investigating itself, that the two are in fact one. Epistemology is no more independent of systematic, directed inquiry than scratching is independent of itching.

This is known as "naturalised" epistemology, and its proponents regard epistemology as a branch of psychology. In the words of the American philosopher W.v.O. Quine:

> Epistemology, or something like it, simply falls into place as a chapter of psychology and hence of natural science. It studies a natural phenomenon, viz., a physical human subject. This human subject is accorded a certain experimentally controlled input—certain patterns of irradiation in assorted frequencies, for instance—and in the fullness of time the subject delivers as output a description of the three-dimensional external world and its history. The relation between the meager input and the torrential output is a relation that we are prompted to study for somewhat the same reasons that always prompted epistemology; namely, in order to see how evidence relates to theory, and in what ways one's theory of nature transcends any available evidence. (*Epistemology Naturalized*, p.82–3)

This assumes a systematic rather than a sentence-by-sentence approach to theories and evidence/justification. The objection to phenomenalism discussed at 10.3, is that there is no evident way that we can reduce the things of the everyday world to sentences about possible perceptual experiences. This goes together with the objections discussed in the previous section to the idea that there is a "given", a way things are that can be stated in such sentences (sentences akin to the "protocol sentences" of 14.2). The next move is to argue that it is only as a complete system that our theory of what there is confronts the evidence that we draw from the senses, and once this step is made any distinction between epistemology as a separate discipline that studies theorising about evidence, and theorising itself, apparently disappears. In place of a distinctive discipline labelled "epistemology", we are left with little more than general considerations about method.

On the other hand, there is still the guiding thought that there must be a connection between inquiring and theorising, and what it is that we inquire and theorise about. Even though naturalising epistemology looks as if it dissolves epistemology, leaving only a descriptive psychological study in its place, the difficulty of doing justice to such concepts as *evidence* and *justification* remains. Whatever else there may be to say about epistemology, there remains a task of *conceptual clarification.*

14.5 Conceptual geography

The central difficulty in contemporary epistemology is a boundary problem. On one side of this boundary is ourselves, minded beings with thoughts and beliefs and, hopefully, knowledge. On the other side of this boundary is what there is, what it is that we have knowledge of. This is, of course, only one way of drawing the boundary. It could be that on this side of the boundary there is only me (solipsism), and other minded beings, if indeed there are any, are inscrutable. (This is the "other minds" problem, of how I can know that there are minds other than my own when I can perceive only other bodies and their emanations.) For the sake of the argument we will assume that a mind can only attain self-consciousness through experiencing other minds, that the boundary is between minded beings and a physical world external to, and independent of, ourselves.

Given this, is it *reasonable* to believe in an external world that is broadly as we perceive it? If we do indeed have knowledge of this external world, can we justify—give reasons for—our knowledge of it?

The difficulty is how to bridge this boundary. Recall Berkeley's remark (10.2), that an idea can be like nothing but an idea. Part of Berkeley's point is that we cannot literally have whatever-it-is in mind; you cannot have what you are thinking about in mind as you can have a pea in a pod or a pearl in an oyster. But given this, what is it that you have in mind when you are knowledgeable about something or other, and how did it get there?

At this point foundationalist projects look particularly attractive, but they fail because there is no clear way of linking the contents of our perceptions with what they are perceptions of. There is no "right" way of perceiving what there is. This ties in with the infinite regress argument discussed in 14.3; justification seems to rely on some solid basis or foundation, but no such foundation is available, so it looks as if we have no justification for what we purportedly know.

One way forward is to weaken the concept of justification, and assert that utility is enough. As long as our beliefs enable us to do what we set out to do, what need have we of justification or even of truth? After all, whether what we believe is true is, arguably, beside the point, if what matters to us is succeeding in what we set out to do. (In other words, truth_{r} is irrelevant to us and truth_{e} just *is* justification, as 3.2.) In a manner of speaking success is its own justification. **Reliabilism**, discussed earlier in this chapter (p.169, above) is just such a non-justificatory approach. What need is there of justification on top of reliability? If a belief-forming process has proved to be reliable in the past, then to the degree it has proven reliable it is reasonable to believe in its deliverances. We can assess the degree of reliability and assign probabilities to outcomes and make suitable adjustments in our belief-forming processes. In a sense this is non-justificatory because it does not use the concept of justification, but this looks like sleight of hand as we could simply claim that the deliverances of a reliable process are *de facto* justified.

At bottom it is hard to see how one can talk at all about knowledge without involving the concept of justification, with having reasons for holding a belief to be true (to be knowledge). If someone says that they know something, we would assume that they can justify—can give reasons for—asserting a knowledge claim. Further, a knowledge claim is *authorising*; if someone says "I know that *p*," then they authorise others to respond appropriately, and they put themselves in a position in which they can be held to account if it turns out that not-*p*. Equally if someone acts on your claim to know that *p*, and not-*p*, then they can offer your assertion as their alibi. It is essential to the concept of knowledge, of expertise, that one who professes knowledge or expertise can be held to account, can be expected to give reasons or justifications for their pronouncements. And if someone cannot give reasons or justifications we would be inclined to deny that they have knowledge even if they happen to be right. (Of course this is the tripartite account of knowledge—knowledge as justified true belief—discussed earlier in 3.2.)

Belief is not similarly authorising. If you say "I believe that *p*" it suggests you have some reason or evidence or justification for *p*, but it is not authorising. While it makes sense to say "I believe that *p*, but I might be wrong," it makes little sense to say "I know that *p*, but I might be wrong."

As for certainty, it might be asked, what difference is there between "I know that *p*" and "I am *certain* that *p*!"? Apart from rhetorical flourish it is hard to see what is added, and one might even wonder if the claim of certainty is suggestive rather of wishful thinking.

14.6 That and how

There is a significant distinction between knowing *that* and knowing *how* that goes together with this boundary problem. We talked (14.2) of believing that *p*, or knowing that *p*, where *p* is sentential; I know that grass is green, that the sky is blue, that seven is greater than six, and so on. Virtually all the theories and approaches we have looked at are concerned with knowing that. What comes after *that* is the natural input to theories of knowledge, because there is not much we can do if all our perceptual/conceptual apparatus yields is "stone" or "blue" or, worse, "this" on being confronted with a something or other.

Sentences are the natural inputs to the inductive and deductive models of argument set out in 6.3. A sentence is the smallest unit of language whose content can be assessed and acted upon: how, after all, should I respond to somebody looking at me and saying "New York"? Put differently, a sentence is true or false, and has consequences either way, whereas a name has no consequences outside of a sentence.

The question that we began with was, how it is that we seem to know more than we have any justification for. If we stick with knowing that, it is not clear how to bridge the boundary between what there is and our knowledge and beliefs, because the latter take the form "that *p*"; they are sentential, they are the sorts of things that are true or false, that we can be asked to justify. To a considerable extent, though, the world just seems to happen to us and does not come packaged in this sort of way. If I say, "there is beer in the fridge," you can assess the truth-value of this and I may be thanked or reproached accordingly. I cannot, though, be held responsible for what I perceive when I open the fridge door. What is there is what is there and I perceive it accordingly.

On the other hand, there is knowing *how*; knowing how to respond to things being thus-and-so, knowing how to go about making them different. This sort of knowledge resists expression in sentential form, and equally resists capture in computational systems. For example, given a finitely long mathematical proof, it can be shown that it is a proof by showing how it can be derived from a set of axioms using valid rules of inference and/or theorems that themselves have been proved from axioms. But there is no obvious way of showing *how* the proof was devised in the first place. In principle the proof could be generated by setting up a programme that generates proofs from sets of axioms, but it still needs someone to devise the axiom sets in the first place.

This relates to the point raised by Meno (2.1), of looking for we know not what. Computers "search" by pattern-matching. If, for example, I ask my computer to search its database of filenames for "/home/clare/Philosophy/Epistemology", it will find the file that contains the text for this book. Miss out one letter and it will not find it. But the terms "search" and "find" treat it *as if* it is human (anthropomorphise it); it simply compares strings of symbols using an algorithm (an effective decision procedure, one that will return an answer in a finite length of time) until it finds a match for the input string or comes to the end of the database, in which case it returns a "file not found" message. (Or simply stops if it has not been programmed to return a suitable message, and sits there mindlessly until I ask it in a language it "understands" what it is doing now.)

By contrast asking a computer to "find" an outfit for a wedding, rather than a file, is not terribly rewarding. By the time all the parameters have been specified, what is wanted is already known, all that is lacking is its location. I know, though, when I see something suitable. But if I say "that is just what I had in mind," it can hardly be that I had it in mind in the same way that a computer has a filename in its memory that it compares to see if there is such a file available to it.

It is not easy to pin down such knowledge. In recent years business consultants have developed and more or less successfully sold the notion of “knowledge management”. The idea is that modern businesses deal in intangibles rather than physical items, so their productive capital is invested in files and reports and databases and the grey goo between their employees’ ears rather than in machine tools and power presses. But knowledge about machine tools and power presses is, in the main, knowledge that, which is why it has proven relatively easy to automate these processes. In modern businesses knowledge is more elusive because most of it is knowledge how; how to respond in given circumstances, how to recognise and act on opportunities, making connections between present and past situations to prevent repeating failed solutions, and so on.

An ideal in knowledge management is to capture and write down what it is that an employee needs to know to do their job. If this could be achieved then the manual would be valuable, not the employee; in principle any employee, defective or not, could be dispensed with and a new one brought in and given the manual, as a part can be changed in a car or a factory staffed by robots.

Even if a manual could be written, though—and millions of pounds later it is far from clear that this is possible for any but the most repetitive of tasks—it is practically impossible to write a manual that covers *all* eventualities. To take a case in point, in the recent past two organisations I have worked for have chosen to redesign their image, involving new stationery. What to do with the old? Keep using it until it is all gone, use it only within the organisation, or discard all of it? Should the manual note, in case of corporate rebranding and consequent new stationery, shred all old headed paper to prevent it walking and being used to create fake job references?

At bottom we are creatures that make *decisions*; we decide in any given set of circumstances what should, or what should not, be the case. And having made decisions we bring them about. When someone says, “I know what the right thing is to do,” it can hardly be that they know in the same sense that they know what the time is or what they had for breakfast. In terms of traditional epistemology, this is a point at which we seem to know more than we have any entitlement or warrant for. To some degree we know how to go about making things better, we know how to exercise autonomy, how to make decisions and act on them. This is not the kind of knowledge that we know how to put into a computer, into any system that operates in terms of knowing that. In this sense theory of knowledge is ultimately a *normative* discipline, it deals not just with how things are and how we can make them different but also with how things should be and how we should go about making them so. What is computable—knowledge that—is, in a manner of speaking, a great deal less interesting than the elusive qualities of minded beings who know how to shape and direct what there is in the pursuit of our goals and purposes.

Glossary

A posteriori *A posteriori* knowledge is knowledge that is drawn from experience and is a function of experience, as opposed to ***a priori*** knowledge. See also **contingency**, 71, 75

A priori *A priori* knowledge is knowledge that is, in some sense, independent of experience. Pinning this "sense" down is difficult. See also **innate ideas**, **applicability** and **necessity**, 70, 75, 106

Anti-Realism See **Realism**, 27

Applicability A problem with ***a priori*** and **necessary** truths, of explaining how they are applicable to the messy **contingent** world we deal with on an everyday basis, 74, 75, 81, 157

Brain in a vat A thought experiment devised by Putnam, that perhaps I am in reality a brain in a vat that is being fed a simulacrum of the "real" world by a mad scientist. A dramatisation of Descartes' malicious demon in the First Meditation, 63, 142

Cogito *Cogito, ergo sum*, Descartes' famous conclusion in the Second Meditation that I think, therefore I am, often referred to as the "cogito", 68

Coherentism Coherentist theories of knowledge treat justification exclusively as a matter of relations between beliefs, with coherence as a criterion for including a given belief within the set, 8, 21, 173, 176, 177

Contingency A contingent state of affairs is one that could be other than it is, for example, that the walls in my room are magnolia; a contingent truth is one that could be false. See also **necessity, logical** and **physical**, 71, 161

Contrary Two items are contrary when they cannot both be true together, but they can both be false together. By contrast the truth of a contradictory implies the falsity of its opposite. A pair of contraries must resemble in some sense; "sick" and "healthy" are contraries, "sick" and "soluble in hydrochloric acid" aren't, 148

Deduction A method used in inferring conclusions from premises according to rules that are truth-preserving; if the premises are true, and the rules of inference are **valid**, then necessarily the conclusion will be true, 84, 102, 116, 124

Defeasible A belief or a theory is defeasible if we accept that the belief or theory may be revised at some future date, 20, 28, 100, 111, 113, 173

Dreaming argument An argument put forward by Descartes in his First Meditation to suggest that our belief in an external world may be delusional, that we may in fact be asleep the whole time and what we think of as our waking life is a continual dream. See also **brain in a vat**, 67, 142

Empiricism Empiricism is the view that knowledge is derived from experience, where experience derives from sensation of what is external to us (outer sense) and what is internal (inner sense), and reflection on that experience. The mind is usually regarded as a "blank slate" that is written on in experience, and powers are attributed to the mind to compare and contrast ideas derived from experience, 40, 82, 86, 102, 104, 111, 113, 173, 176

Essence The defining property or quality of something, what makes it distinctively what it is, from the Latin *essentia* meaning the being of something. See also **accidental properties**, 33, 47, 55, 61, 62, 84, 97–99, 111, 116, 151

Essence, nominal Nominal essence derives from a Medieval notion, that there isn't something (a **form** or **essence**) that similar things have in common, what they have in common is simply our preparedness to apply a common label to them, 94, 97, 99

Essence, real The real essence of something is what it is in itself, its defining characteristic. In Locke this may be its real and unknowable essence, or its real essence insofar as it can be discovered by us. Real essence is thus opposed to **nominal essence**, 94, 97, 99, 102

Falsificationism An approach to scientific method proposed by Popper, that the mark of a properly scientific theory is not more and more confirming instances (see also **induction**) but that we have a clear idea of what would count as a disproof or a disconfirming observation or experiment, 130

Form The *form* of something is what makes it what it is, its property or quality. "Forms" are usually Platonic forms, associated with Plato's theory of forms. See also **essence**. In Aristotle's philosophy things (substances) are a unity of form and **matter**, 33, 61, 75, 102, 116

Foundationalism Foundationalist theories of knowledge begin with suitably justified basic beliefs and then build on these, with more complex beliefs supported by the basic beliefs (dependence and justification are one-directional), 8, 90, 173, 176, 177

Hypothesis A hypothesis is a thesis arrived at on the basis of observation and, perhaps, experiment, that is put forward for further testing and experimentation. A hypothesis is **defeasible**, 81, 106, 172

Idealism Idealism is the view that there are only minds and their ideas, that all there is is mind-stuff. This doesn't preclude the idea that there is something external to me, only it will be a mind or minds. See also **realism** and **phenomenalism**, 126, 134, 176

Induction A method of reasoning that relies on accumulating confirming observations as a basis for generalisations, of the form this case is x, this case is x, ..., therefore all such cases in the future will be x. An inductive generalisation is **defeasible** because it may be modified in the light of a counter-example or may be held onto until disconfirming evidence forces us to revise our theory of whatever it is, 81, 83, 84, 102, 130, 174

Innate ideas Innate ideas are ideas that we have in mind prior to any experience of the world, ideas that, according to Descartes, are implanted in us by God. A way of accounting for **necessity**, 48, 56, 60, 70, 74, 80, 87, 106

Metaphysics The branch of philosophy that deals with **substance**, with "being" in its widest and most general sense, 45

Necessary and sufficient conditions An analysis of a concept in terms of necessary and sufficient conditions aims at an identity statement in which the right-hand side expresses conditions that must be met for a concept to be instanced. What meets the conditions set out on the right-hand side is an instance of the concept (sufficiency) and isn't something else (is necessarily an instance of this concept), 26

Necessity Necessity is a property of things that if they are necessary, then they could not have not existed and they could not have been different from how they are. Similarly necessary truths are truths that could not have been false, 71, 102, 116, 157

Necessity, logical Logical necessity applies to truths that cannot be false because their opposites are self-contradictory; so $2 + 2 = 4$ is necessarily true because $2 + 2 \neq 4$ is contradictory. This gets difficult if laws of nature are ***a priori*** and consequently seem to be logically necessary, 157

Necessity, physical If laws of nature aren't **logically** necessary, then perhaps they are **physically** necessary—that is, true in all possible worlds like ours but not self-contradictory, 157

Omniscience problem The omniscience problem is a problem for **rationalism**, because the notion of **innate ideas** can all too easily attribute to us a degree of knowledge that can only plausibly be ascribed to an omniscient being, 58, 102, 104

Phenomenalism The view that things have no significant existence beyond our perceptions of them, that things are "permanent possibilities of sensation" (J.S. Mill), 134

Projectibility An idea proposed by the American philosopher Nelson Goodman, in response to Hume's approach to induction, that perhaps we can rely on some predicates/law-like relations because they can be projected into the future, 149, 150

Proposition By a *proposition* is meant the content of what a sentence says, or the content of a thought, stripped of **contingent** accretions due to language, 170

Qualities, primary The primary qualities of something are typically shape, duration, position, number, and so on, the qualities of a body that are addressed by mathematical physics. They are the qualities that a body has irrespective of us, its *real* qualities. See also **realism**, **essence**, 65, 66, 94, 120

Qualities, secondary A body's secondary qualities are those we perceive it in terms of, typically colour, smell, taste, hardness or roughness. These qualities are not in the body, they are caused (in us) by its **primary qualities**, 65, 66, 94, 120

Rationalism Rationalism is the view that some, or perhaps all, that there is to know can be derived from reflecting on the contents of our minds alone, without reference to an external world. See also **innate ideas**, **a priori** and **necessity**, 39, 80, 102, 104, 106, 108, 113, 124, 176

Realism Realism is the view that what we are in perceptual or cognitive contact with is independent from us and is as it is whether there are human beings or not. The generic opposite is **anti-realism**, including **idealism** and **phenomenalism**, 27, 91, 133

Realism, naïve The view that we experience things directly, without intermediaries, 92, 133

Realism, representative Representative realism holds that what we experience in perception is what there is but somehow mediated, by ideas of things in terms of their **secondary qualities**. See also **veil of perception**, 92, 96, 120, 133

Reliabilism A non-justificatory approach to knowledge, that what it is to know something-or-other is to be reliable about whatever-it-is, 169, 181

Scepticism Scepticism is bound up with justification and knowledge-claims, with whether there are rational grounds for doubting that we can ever attain knowledge. Scepticism can be academic or dogmatic (we can never attain knowledge) or Pyrrhonian or mitigated (we cannot know that we can ever attain knowledge), 137

Substance "Substance" is the basic stuff of what there is, the underlying substrate of the things we experience. Substance is usually thought of as a metaphysical rather than as a scientific concept. A substance, particularly in Aristotle, can also be the sort of thing that can be the subject of a sentence, that has an independent existence (as opposed to a predicate or **concept**), 97, 120, 122

Truth It is impossible to give a succinct statement of truth and falsity, but please read on, 26

Validity A valid argument is one constructed according to rules of inference that are truth-preserving, that is, if the premises are true and the rules are valid then the conclusion is **necessarily** true, 106

Veil of perception A way of expressing the idea that we can only ever experience what there is in terms of **secondary qualities**, that things as they are in themselves (in terms of their **primary qualities**) are inaccessible to us, 69, 96

Verificationism An approach to justifying theories, of stating what would count as a confirmation of a theory and then piling up confirming instances, 130, 170

Vortex theory Descartes' explanation for the circular orbits of planets, that they are carried round a star on a series of concentric vortexes (assuming the Aristotelian claim that motion can only occur through contact), 109, 110, 112, 132

Bibliography

Aristotle: 1955, *On Coming-to-be and Passing-away*, Loeb Classical Library, Harvard University Press, Cambridge, MA, 1955. (trans. E.S. Forster).

Aristotle: 1996, *Physics*, Oxford University Press, Oxford, 1996. (trans. R. Waterfield).

Aristotle: 2004, *Metaphysics*, Penguin Books, London, 2004. (trans. H. Lawson-Tancred).

Ayer, A. J.: 1936, *Language, Truth and Logic*, 2nd edn, Penguin Books, London, 1946.

Berkeley, G.: 1734a, *Three Dialogues between Hylas and Philonous*, 3rd edn, Oxford University Press, Oxford, 1998.

Berkeley, G.: 1734b, *A Treatise Concerning the Principles of Human Knowledge*, 2nd edn, Oxford University Press, Oxford, 1998.

Boyle, R.: 1686, *A Free Enquiry into the Vulgarly Received Notion of Nature*, Cambridge University Press, Cambridge, 1996.

Carnap, R.: 1935, *Philosophy and Logical Syntax, Psyche Miniatures no.70 General Series*, Kegan Paul Trench Trubner & Co., London, 1935.

Descartes, R.: 1637, *The Geometry of René Descartes*, Dover Publications Inc., New York, 1954.

Descartes, R.: 1642, 'Meditations on First Philosophy, with Objections and Replies', *in* J. Cottingham, R. Stoothoff and D. Murdoch (eds), *The Philosophical Writings of Descartes Vol.II*, Cambridge University Press, Cambridge, 1984.

Descartes, R.: 1644, 'The Principles of Philosophy', *in* J. Cottingham, R. Stoothoff and D. Murdoch (eds), *The Philosophical Writings of Descartes Vol.I*, Cambridge University Press, Cambridge, 1985.

Descartes, R.: 1991, *The Philosophical Writings of Descartes Vol.III, The Correspondence*, Cambridge University Press, Cambridge, 1991.

Einstein, A.: 1954, *Relativity, The Special and the General Theory*, 15th edn, Routledge, London and New York, 2001.

Euclid: 1956, *The Thirteen Books of Euclid's Elements*, 2nd edn, Dover Publications Inc., New York, 1956.

Gettier, E.: 1963, 'Is Justified True Belief Knowledge?', *in* S. Bernecker and F. Dretske (eds), *Knowledge, Readings in Contemporary Epistemology*, Oxford University Press, Oxford, 2000.

Hall, A. R.: 1981, *From Galileo to Newton*, Dover Publications Inc., New York, 1981.

Hume, D.: 1739, *A Treatise of Human Nature*, 2nd edn, Clarendon Press, Oxford, 1992.

Hume, D.: 1777, *Enquiries Concerning Human Understanding and Concerning the Principles of Morals*, 3rd edn, Clarendon Press, Oxford, 1979.

Kant, I.: 1781/7, *Critique of Pure Reason*, Cambridge University Press, Cambridge, 2000.

Kant, I.: 1783, *Prolegomena to Any Future Metaphysics That Will Be Able to Come Forward as a Science*, Cambridge University Press, Cambridge, 1997.

Locke, J.: 1700, *An Essay Concerning Human Understanding*, 4th edn, Clarendon Press, Oxford, 1979.

Mill, J. S.: 1867, *An Examination of Sir William Hamilton's Philosophy*, in *Collected Works of J.S. Mill Vol.IX*, 3rd edn, University of Toronto Press/Routledge and Kegan Paul, Toronto/London, 1979.

Moore, G. E.: 1939, 'Proof of An External World', *in* T. Baldwin (ed.), *G.E. Moore, Selected Writings*, Routledge, London, 1993.

Moore, G. E.: 1941, 'Certainty', *in* T. Baldwin (ed.), *G.E. Moore, Selected Writings*, Routledge, London, 1993.

Newton, I.: 1726, *The Principia, Mathematical Principles of Natural Philosophy*, 3rd edn, University of California Press, Berkeley, 1999. Trans. Cohen & Whitman.

Newton, I.: 1729, *The Mathematical Principles of Natural Philosophy*, 3rd edn, Dawsons of Pall Mall, London, 1968. Trans. Motte.

Newton, I.: 1730, *Opticks*, 4th edn, Dover Publications Inc., New York, 1979.

Newton, I.: 2004, *Isaac Newton, Philosophical Writings,* edited A. Janiak, Cambridge University Press, Cambridge, 2004.

Plato: 1953a, *Meno*, in *The Dialogues of Plato translated into English by B. Jowett, Vol.I*, 4th edn, Clarendon Press, Oxford, 1953.

Plato: 1953b, *Phaedo*, in *The Dialogues of Plato translated into English by B. Jowett, Vol.I*, 4th edn, Clarendon Press, Oxford, 1953.

Plato: 1953c, *The Republic*, in *The Dialogues of Plato translated into English by B. Jowett, Vol.II*, 4th edn, Clarendon Press, Oxford, 1953.

Plato: 1953d, *Theaetetus*, in *The Dialogues of Plato translated into English by B. Jowett, Vol.III*, 4th edn, Clarendon Press, Oxford, 1953.

Quine, W. v. O.: 1969, 'Epistemology Naturalized', in *Ontological Relativity and other essays*, Columbia University Press, New York, 1969.

Russell, B.: 1906, 'On the Nature of Truth', *Proceedings of the Aristotelian Society* **vol.7**.

Index

Printed in the United States
147084LV00002B/3/P